The Body of Evidence

How the Human Body Refutes Intelligent Design

Rosa Rubicondior

The Body of Evidence

Cover Design: AI-generate (ChatGPT4o)

Third Party Copyright.

Third party copyright is acknowledged for work not currently in the public domain, reproduced here for criticism and analysis under intellectual rights fair use regulations.

In the opinion of the author, the minimum necessary for effective criticism and analysis whilst retaining the original context, has been reproduced in this work.

ISBN: 979-8304445078

© 2024 Rosa Rubicondior.
All rights reserved.

The Body of Evidence

Contents

Introduction 7
   Irreducible Complexity...9
      The Irreducibly Complex Eye Fallacy 10
      The Coagulation Cascade Fallacy 11
   The Argument from Perfection 14
   Fine-Tuning and Purpose 17
   The Beauty and Aesthetics of the Body 19
   The "Just-Right" Argument...21
   Rejection of Evolutionary "Explanations for Flaws" 27
Part I: Flaws in Basic Architecture..33
   From the Top Down 35
      The face and skull..35
      Wisdom teeth..38
      Choking hazard...39
      The cervical vertebrae 40
      The lumbar vertebrae..41
      The pelvis and pelvic floor..42
      External testes...43
      The hip joints..43
      The knees..44
      The ankles and feet...44
Part II: Internal Weaknesses...47
   The Needlessly Complicated Digestive Tract 49
      Hiatus hernia...49

    Gastric ulcers 50

    Gall stones and risk of pancreatitis 51

    Diabetes. 52

    Paralytic ileus 55

    The appendix 59

    Ulcerative colitis, etc. 60

    Irritable Bowel Syndrome 64

    Diverticula and Diverticulitis 68

    Haemorrhoids 70

The Delicate Heart and Circulatory System 75

    Blood clotting and thromboses 75

    Electrical control of the heart and arrhythmias 78

    The pulmonary system 85

    Varicose veins 92

    Coronary circulation and heart attacks 95

The reproductive systems 101

    The Male Reproductive System 101

        The vas deferens. 101

        Testicular Vulnerability. 102

        Prostate Gland Issues. 102

    The Female Reproductive System 103

        Birth canal and pelvic evolution 103

        Ectopic Pregnancies. 103

        Placenta previa. 103

        Pre-eclampsia. 105

        Menstruation. 107

Contents

Hyperemesis gravidarum (severe morning sickness)...109
Menopause...112
Reproductive Cancers...113
Part III: Inefficient or Wasteful Systems...115
The Energy Hogging Brain...117
    Coning...120
    Concussion...121
    Mental health disorders...122
    Epilepsy...125
The Immune System's Paradox...129
The Wasteful Reproductive Strategy...133
Part IV: Evolution Explains it All...139
Vestigial Organs, Atavisms and Exaptation...141
  Vestiges of ancient ancestry...141
    Coccyx (Tailbone):...141
    Wisdom Teeth...142
    Plica Semilunaris (Third Eyelid)...142
    Erector Pili Muscles (Goosebumps)...143
    Auricular Muscles (Ear Movement)...143
    The Preauricular Pit...144
  Atavistic traits...145
    Tail Formation...146
    Extra Nipples (Polythelia)...146
    Dense Body Hair (Hypertrichosis)...146
  Exaptation of redundant DNA...147
    Genes from Endogenous Retroviruses (ERVs)...148

- Pseudogenes Acting as Regulatory Elements 148
- Gene Duplication and Divergence 149
- Transposable Elements as Regulatory Elements 149
- FOXP2 and Speech 150
- Human-Specific Enhancers 150
- Vitamin C Synthesis Gene (GULO) 150
- Path Dependence in Evolution 151
  - The Pharyngeal Arches and Craniofacial Development 152
  - The Human Spine 152
  - The Human Eye 152
  - The Human Pelvis and Childbirth 153
  - The Vas Deferens in Males 153
- Conclusion 155
- Appendix 163
  - Creationist claims rebutted 165
    - The Human Eye 165
    - Blood Clotting Cascade 166
    - The Human Brain 166
    - Human Reproductive System 167
    - The Appendix 168
    - The Immune System 168
    - Vestigial Structures 169
    - The Heart and Circulatory System 169
    - Human DNA 170
- References 171

Contents

Index 175
Other Books by Rosa Rubicondior 205

The Body of Evidence

Introduction

To a creationist, the human body represents the supreme achievement of their putative designer god, standing at the pinnacle of designed perfection. Even many theists who accept that human beings are the result of an evolutionary process, believe that God (it is always the locally-popular god) intervened and guided the process to ensure humans were the ultimate result.

It follows from that false argument that the eventual evolution of humans was the entire *raisons d'etre* of the Universe itself, since there is a chain of causality leading from the Big Bang to the evolution of humans, in which a designer god was supposedly ever-present, ensuring a sun of the right size formed with a planetary system, one planet of which was 'perfect' for life to exist on, etc, etc, as a precondition for evolving humans.

I explained how, following the Big Bang, the operation of the basic laws of chemistry and physics led to the evolution of everything, including humans, human cultures, and eventually each and every one of us, without needing a god anywhere in the explanation, in *What Makes You So Special? From the Big Bag to You* (1), but creationists insist their particular god was involved at every step along the way.

It is, of course, nothing more substantial than arrogant anthropocentrism – exactly the belief that in earlier times argued that Earth stood at the center of the Universe.

The assumption is always that the current situation was the intended situation; that there was intent behind everything, and that intent was to produce what we have today; not what we had 100 years ago, 1000 years ago, or 10,000 years ago, but what we have today; though no doubt people alive then believe their situation was the purpose of it all.

How thoughtful of the locally-popular god to provide caves to live in and flint to make spears from the better to hunt mammoths, that were obviously there to provide food, skins and bone.

Arguments for a designer god are invariably arguments from ignorant incredulity, which assume that because they don't know something, it is unknowable by science. Even when they are right and science doesn't yet have a full explanation for something, that fact in itself is not an argument for creationism; creationists need to provide evidence for their beliefs because they don't win against science by default. Creationism, like any other hypothesis, needs to provide scientific evidence for itself or it can be dismissed as just another evidence-free guess.

So, just what is the creationist argument for the designed perfection of the human body and how does that compare with what science shows us to be the facts?

Creationists argument almost invariably depend on tactic in lieu of factual arguments such as :

- Irreducible complexity
- The argument from perfection
- Fine tuning and purpose
- Beauty and aesthetics of the body
- The 'just right' argument
- Rejection of evolutionary explanation for flaws

To take these one at a time:

Introduction

Irreducible Complexity.

A flawed notion introduced by Professor Michael J. Behe, in which he argues that if a structure or process needs all its component parts in place before is serves any useful purpose, it could not have evolved by slow Darwinian 'one-small-step-at-a-time' evolution.

Professor Behe is a professor of biochemistry at Le High University who's views and conclusions have been publicly and unanimously repudiates by his colleagues in the Le High Biochemistry Department.

There are a number of flaws in this argument. Firstly, it ignores the result of exaptation of redundant or duplicated structure originally evolved for a different purpose, for example, the components of the bacterial flagellum that Behe used as his example in *Darwin's Black Box: The Biochemical Challenge to Evolution* (2) is now known to have existed in the Type III Secretory System (T3SS).

Secondly, it is a false dichotomy fallacy in which the argument is presented as a choice limited to Darwinian evolution or intelligent design, with no other possibilities allowed, even if evolution had really been excluded.

Thirdly, it is a 'god of the gaps' argument in which a gap in scientific knowledge, real, imaginary, or manufactured, is filled with a god specially designed to fit the gap.

The latter two flaws in the 'irreducible complexity' argument are designed to appeal to a parochial and culturally chauvinistic audience which can be depended on to supply the locally-popular god the presupposition needs without having to define it, explain it or provide any evidence for it, so appealing to religious fundamentalists while pretending not to be a religious argument.

The Irreducibly Complex Eye Fallacy

In reference to the human body specifically, creationists often claim the eye as an example of an irreducibly complex organ, usually accompanied by a quote mine from Charles Darwin's *On The Origin of Species* (3), in which, taken out of context, he appears to say the eye can't be explained, whereas, in context, he is simply giving this as an example of something that can't be explained without his Theory of Evolution. In fact, as Darwin suggests, there exists a whole range of structures from a small group of light-sensitive cells to the mammalian eye, complete with retina, lens and muscles, showing how the eye could indeed have evolved via a progressing sequence of increasing complexity. There are also many different eyes in nature from the molluscan eye of the cephalopods to the compound eyes of the arthropods. In short, eyes not only evolved but did so several times.

The problem for creationists wishing to cite the eye as an example of how humans sit at the pinnacle of created perfection is not only that its evolution is fully explainable with examples without needing a god in the explanation, but it is also far from being the most perfect of eyes.

There are many vertebrates with a much more efficient eye than the human eye, for example, a stooping peregrine falcon travelling at over 200 mph, has a transparent nictitating membrane to cover the eye to prevent it being damaged by dust and flying insects while still focussing on its prey. Anyone who has ridden a motorcycle without goggles can testify to what a boon it would be to have just such a built-in pair of goggles.

The sparrowhawk can fly at top speed through woods in pursuit of small birds, because its 'frame rate' enables the branches to stay in focus while moving at speeds at which, to a human, everything would appear as a blur. Lastly, a golden eagle's visual acuity is such that is able resolve fine details the equivalent of the typescript of a standard newspaper at a distance of a mile.

The Coagulation Cascade Fallacy

A second argument commonly deployed by creationists is that of the coagulation cascade, despite the fact that it has been explained as the result of an evolutionary process. This argument, like many creationist arguments, relies on misrepresenting the science to scientifically illiterate, and often wilfully so, creationists, looking for confirmation of pre-exiting bias.

The coagulation cascade is a complex process involving a sequence of biochemical reactions that prevent excessive bleeding by forming a stable blood clot. It involves multiple proteins, enzymes, and cofactors which are primarily produced by the liver and circulate in the blood in inactive forms. They are normally given Roman numerals for convenience.

When an injury occurs, these are activated in a stepwise manner, initiated by factors released by damaged cells causing platelets[1] to stick together and initiate the next step in the process. The cascade sequence is:

1. Primary Haemostasis: triggered by blood vessel injury. This exposing collagen and tissue factor (TF) This causes platelets to adhere to the site via von Willebrand factor and become activated, releasing signalling molecules to recruit more platelets and form a temporary "platelet plug."

---

[1] Platelets are cell fragments produced by very large, megakaryocyte cells in bone marrow

The Body of Evidence

2. Initiation of the Coagulation Cascade: Tissue factor (TF) combines with Factor VII to form the TF-VIIa complex. This complex activates Factor X to Factor Xa.

3. Amplification via the Intrinsic Pathway: Exposed negatively charged surfaces (like collagen) activate

   - Factor XII to Factor XIIa.
   - Factor XIIa activates Factor XI to Factor XIa.
   - Factor XIa activates Factor IX to Factor IXa.
   - Factor IXa, with its cofactor Factor VIIIa, activates more Factor X to Xa.

4. Thrombin Generation: Factor Xa, with its cofactor Factor Va, converts prothrombin (Factor II) to thrombin (Factor IIa). Thrombin amplifies the cascade by activating Factors V, VIII, and XI, creating a positive feedback loop.

5. Fibrin Clot Formation: Thrombin converts fibrinogen into fibrin, which polymerizes to form a mesh-like structure. Factor XIIIa cross-links fibrin strands, stabilizing the clot.

6. Termination and Regulation: Natural anticoagulants (e.g., antithrombin, protein C, and protein S) prevent over-clotting and the fibrinolytic system dissolves the clot once the injury heals.

The coagulation cascade has evolved over hundreds of millions of years during which existing proteins were co-opted and modified to create new functions.

That it evolved out of an earlier system is evidence by the fact that simpler systems exist in earlier invertebrates such as

Introduction

arthropods and molluscs which rely on aggregation of cells at the site of the injury or simpler protease cascades for wound sealing.

Some of the clotting factors are structurally similar and evolved by gene duplication and repurposing. Factors like fibrinogen and thrombin are also the product of gene duplication, having evolved from ancestral enzymes involved in other proteolytic processes.

In vertebrates, the process has evolved a belt and braces approach as two pathways, the intrinsic pathway initiated by contact with exposed proteins such as collagen at the injury site, and the extrinsic pathway initiated by tissue factors released by damaged cells.

Finally, in mammals there were further refinements, particularly to factors VIII and IX which became more efficient and the evolution of the regulatory mechanisms (e.g., protein C) which, in another example of an additional layer of complexity to compensate for a suboptimal process, to reduce the risk of excessive clotting, which could cause thrombosis.

The evidence that the coagulation cascade is an evolved process can be summarised as:

> Homology Between Clotting Factors: Many clotting factors are structurally similar, indicating they arose from gene duplication and divergence. For example: Factor X, Factor IX, and protein C share a common genetic ancestor.

> Comparative Genomics: Simpler clotting cascades exist in jawless fish (e.g., lampreys), while cartilaginous fish (e.g., sharks) have a more developed system and amphibians and reptiles show intermediate stages of the cascade compared to mammals.

Experimental Evidence: Researchers have shown that specific clotting factors can be removed or altered leaving the process still functional but less efficient, showing how much of the evolution of the cascade has been with improving efficiency rather than providing function. This of course gives the lie to claims that the process is irreducibly complex so could not have evolved by a Darwinian evolutionary process.

The Argument from Perfection

The subjective argument from perfection depends on a circular argument in which the human body is arbitrarily defined as perfect. It is far from it, of course, which is why it exists in a wide range of shapes and sizes, heights, strengths and speeds and stamina for running and jumping.

As well as anatomical difference, human also vary in their DNA, their physiology, the efficiency of their immune systems. This is why clinical trials are ideally conducted on a wide range of people from different ethnic and socioeconomic backgrounds, preferably across several different countries, because researcher know that there are significant difference between populations, so the results of a trial conducted on a single population might not be applicable to all humans.

These of course are the variations on which natural selection acts to drive evolution, but they give the lie to an arbitrary definition of perfection since perfection would not have significant variations, and what intelligent designer is going to design lots of subtle variations on a basic theme?

The circularity of the claim that anything designed by God must be perfect, therefore the human body must be perfect, therefore the human body was designed by God, is a good example of the intellectually bankrupt creationist reasoning.

Introduction

Two examples of this design 'perfection' are normally cited, the first being the human brain which is claimed to be the most complex and efficient organ in existence – which begs the question, how do they know this? We do not know if there are even more complex organs in beings on other worlds in the Universe, but there is no denying that the human brain is the most complex vertebrate brain – though not the largest – that distinction, amongst terrestrial mammals, belongs to the African Elephant.

The adult African elephant has a brain which weighs 4.7-5.5 Kg making it about four times the weight of a human brain. It also has about 257 billion neurones, i.e. about three times as many as humans. Their brain is structurally similar to that of humans, with large cerebella which are involved with movement and coordination, large temporal lobes which are involved with memory, and large hippocampi which are involved with emotions.

The Bottlenose dolphin also has a larger brain that humans, weighing about 1.6 Kg, compared to the human 1.3-1.4 Kg, however it differs structurally in a few significant ways: its neocortex is smaller than that of humans, comprising only 40% of the mass of the brain against 80% in humans, but a dolphins limbic system, which controls emotions, is more complex than the human limbic system. Neither dolphins nor any other toothed whales have olfactory lobes which suggests they don't have a sense of smell.

However, humans', elephants' and bottlenose dolphins' brains are all dwarfed by that of a sperm whale which weighs in at about 9 Kg. - over six times larger than a human brain. Sperm whales are probably the most intelligent creatures in the ocean, having complex social structures and the ability to communicate over enormous distances with complex sounds suggestive of a language. Moreover, sperm whales' brains have changed little since the cetaceans first evolved about 55 million years ago; humans only acquired their large brains some 400,000 years ago. The largest hominin brain was probably that of Neanderthals (although we don't have a Denisovan cranium to measure), which was slightly larger than that of *Homo sapiens*, but may have been arranged differently.

The point of these comparisons is to show that brain size and complexity have evolved differently in widely different species as a result of environmental pressures which have produced brains which fit the species for survival and reproduction in their respective environments. One feature these species, all have in common is organised social structures, so much of the brain may be devoted to communication and empathy.

In addition to our large brains, we also have an opposable thumb which means we can touch the tip of our thumb with the tip of each of our other fingers – the only vertebrate that can do that. This gives us the ability to manipulate objects and use tools with precision. To a creationist having a unique ability is often waved around as evidence that humans are a special creation, seemingly oblivious to the fact that every species has unique characteristics, which is why they are classified as different species.

During human evolution, having descended from the trees to a terrestrial existence, an change which included the evolution of bipedalism, our hands were freed from their former function of grasping branches, where they need to function more like a hook than a tool-grasping hand.

One archaic hominin, *Australopithecus sediba*, from South Africa, had a hand which differed little from that of a modern *H. sapiens* hand, in addition to lower limbs which were almost identical to those of *H. sapiens*, so the evolutionary process of

Introduction

changing the proto-human hand into a tool-grasping hand had already occurred at least once by about 1.9 million years ago.

The evolution of the hand and the human brain are examples of co-evolution within a species as synergy between different traits produces additional selection pressure to improve them both simultaneously.

The same process applies to the processes of communication, memory and learning, especially learning the social skills needed to function as an adult in human society, and a growing brain led to another feature of humans which has proved fortuitous – early birth to deliver a large head through the birth canal made curved by the process of bipedalism, which results in a long childhood during which physical and mental skills can be learned, creating an environment for memetic, or cultural evolution.

Fine-Tuning and Purpose

The argument that humans are 'fine-tuned' for survival, reproduction and environmental interaction – all of which are predictable from the first principles of evolutionary theory – being used to justify the claim that this is evidence of design, is an example of presuppositional circular reasoning, and an example of where Occam's Razor[2] should be applied to the argument to pare away extraneous and unnecessary entities. The Theory of evolution provides the most vicarious explanation, so trying to insert a god (always the locally-popular god) into the answer, simply to satisfy a deep psychological need to include it, is to multiply entities needlessly.

---

[2] Occam's Razor is a philosophical devise to ensure the least complex answer as the most likely explanation for anything. Named after William of Occam (or Oakham) who reasoned that one should not multiply entities needlessly since the most vicarious explanation is probably the right one.

The Body of Evidence

The claim is often made that the structure of DNA, which is often cited as a "blueprint" for creating, in this case, a human, must have been intelligently designed.

This of course, ignores that fact that DNA is not a blueprint that some mysterious process reads as a set of instructions on how to assemble a human baby from a list of parts, like assembling a Bili bookcase from Ikea (other bookcases are available), but a data store in which templates for proteins are contained in sequences which code for particular amino acids. These are transcribed into RNA using the enzyme, transcriptase which recognised other codes within the DNA such as start and stop. These DNA sequences can be activated and deactivated using small non-coding RNA sequences which add or remove methyl groups to certain sequences in the RNA which prevent transcriptase transcribing them. Several key clusters of genes can be turned on or off with homeobox (hox) genes which instruct a mass of cells in the developing embryo to 'grow a limb', 'make and eye', etc.

How the information stored in DNA is used to make a new individual is a process that has been evolving since the first prokaryotes[3] used DNA as RNA's data store, and it was given new impetus when the symbiotic associations of prokaryote cells we call eukaryote[4] cells first formed multicellular organisms

---

[3] A prokaryote is a single-cell organism whose cell lacks a nucleus and other membrane-bound organelles. The word prokaryote comes from the Ancient Greek πρό (pró), meaning 'before', and κάρυον (káruon), meaning 'nut' or 'kernel'. Prokaryotes are divided into bacteria and archaea based on structural differences in their cell membranes. (Wikipedia)

[4] The Eukaryotes constitute the domain of Eukaryota or Eukarya, organisms whose cells have a membrane-bound nucleus. All animals, plants, fungi, seaweeds, and many unicellular organisms are eukaryotes.(Wikipedia)

Introduction

around 1.7 billion years ago, so they have had a long time to get better at it.

But of course, the best argument against the notion of intelligent design is that of variation within a species. Your genome is not the same as mine or any of your siblings or close relatives and yet, by and large, it does the same thing, albeit with variations in the outcome – which is why you don't look like me and, with the exception of some identical twins, you can tell brothers and sisters apart. Where is the intelligence in creating tens of billions of variations on a basic theme?

Occam's Razor tells us to pare away those needless entities and go with the most vicarious explanation – mindless, unplanned evolution.

The argument that humans have a purpose is mere anthropocentric arrogance. If anything can be described as a purpose for our existence it is to produce the next generation and raise them to adulthood so they can produce children in their turn. The basic purpose of reproducing is no different now than it was when the first self-replicating molecule made the first copy of itself. Beyond that, our life has the purpose we choose to give it. It is not for the followers of one or other religion to assign that purpose for us. To try to abrogate that right is arrogance of the first order.

The Beauty and Aesthetics of the Body

What, all of them?

The Body of Evidence

The more interesting question is not the childish teleological[5] one of who or what created the perceived aesthetic beauty of the human body, but why do we perceive it as beautiful?

And moreover, why is the idea of beauty, subject to cultural variation?

In my former role as an information systems manager, one of my department's functions was to produce Microsoft Access solutions for information reporting and retrieval. One of my staff was an extremely talented programmer from Southern India whose designs I sometimes needed to tone down from the bright colours that doubtless would have appealed to an Indian user, but which would not have appealed to the more subdued aesthetics of the, mostly, Europeans who were to use them.

Aesthetics are culturally subjective and not something that can be objectively measured except in the most general terms, such a symmetry.

However, given that humans are bilaterians – something they share with every one of their ancestors going back to before the Cambrian when the first chordates appear, it would be astonishing if humans displayed anything other than bilateral symmetry., derived ultimately from the way the fertilised zygote divides initially into two identical cells differing only in their juxtaposition with regard to the other one of the pair.

---

[5] Teleology is the assumption of design. It tends to predominate in the thinking of young children who see the world in terms of sentient cause for every phenomenon – a view that imagines every object including elementary particles are sentient to the degree necessary to understand and obey instructions. When teleological thinking is retained into adulthood, it is the usual cause of both creationism and conspiracism which often occur together, so science which contradicts creationism is dismissed as part of a worldwide conspiracy by scientists.

Introduction

Additionally, our faces, by which we recognise parents, friends, relatives and co-workers are symmetrical with eyes placed about one third of the way down from the hairline, and our mouth is about one third of the way up from the chin, so much of our aesthetic appreciation comes from observing, as any professional artist or photographer will be aware, the 'rule of thirds' where the focal points are on lines dividing the image into thirds.

In other words, there are many aspects to our aesthetics, some deeply embedded in cultures and some to our development as children, but symmetry itself is deeply embedded in out evolutionary history.

Again, no gods required in the explanation because they provide nothing that can't come from psychology, and memetic and genetic evolution.

The "Just-Right" Argument

This piece of circular reasoning comes straight out of the teleologist's playbook. It argues that the body must have been designed because it is 'perfectly' suited to its environment. Try telling that to an someone who finds themselves outside in a blizzard with no coat on.

Humans are only anywhere near perfectly suited to living in East Africa or tropical areas, and then well away from parasite vectors like mosquitos and tsetse flies, and with enough shade to shelter from the intense heat of summer at midday. For almost everywhere else in the world, for those places that are not underwater, we need at least clothes and shelter and sometimes special equipment to survive more than a few hours.

An example often cited is that of upright walking – something we shared with our two closest living relatives, the chimpanzee and the bonobo, both of which can walk upright especially when carrying a load. It is something we almost certainly inherited

The Body of Evidence

from our last Australopithecine ancestor, *Australopithecus afarensis* or a close relative.

The primary cause of the evolution of bipedalism in the East and South African Australopithecines was probably the same as the cause of our divergence from the common ancestor of hominins and the chimpanzees – climate change causing the forest to become savannah with scattered trees.

This meant an arboreal existence was no longer an option and early hominins had to adapt to a hunter-gatherer lifestyle, which meant covering a wide area of land as energy-efficiently as possible so there was strong selection pressure to become fully bipedal. This includes completing the migration of the foramen magnum forward in the skull, which had begun in our simian ancestors, to place the head more centrally on top of the spinal column, and for the muzzle to recede under the expanding brain case bringing the eyes into a forward-looking position when the trunk is upright.

The upright posture entailed two 'unnatural' curves in the spine – the lumbar curve and the cervical curve. This also meant a remodelling of the pelvis to allow the legs to hang straight down and to act as a bowl to contain the organs of the lower abdomen which had previously been suspended beneath a spinal column with a single curve, the remnant of which is now seen only in our thoracis vertebrae.

Although these changes enabled our archaic ancestors to run and hunt efficiently, and the later invention of stone tools and hunting equipment meant we could hunt and run-down large game, which combined with the invention of fire and cooking, gave us the necessary nutrition to grow a large brain, they were not without costs. As I will show later, each of these major changes to our skeleton, from the pelvis to the spinal column and the skull was to lead to problems which plague is still today, and

Introduction

which give the lie to notions of intelligent design. We are the result of utilitarian compromises and trade-offs between competing natural selectors in our environment.

As we evolved away from the trees that formerly gave us a degree of safety, we were never going to succeed in the African savannah as individual apes, no matter how clever we were or how sophisticated our tools became. A solitary human armed only with a spear is no match for a lion and is almost defenceless against a leopard dropping from a tree. Bringing down a kudu bull or even a small gazelle takes a band of hunters, not a single person with a spear, because it is a team effort.

So, there was pressure on us to develop communication and tracking skills and to accept a hierarchical social structure with a team leader who coordinates the hunt and uses skills taught to him by peers and authority figures in the tribe.

Consider for a moment the cognitive skills in being able to 'read' the narrative in animal tracks. t We can tell which direction the animals was walking in, which tells us where it came from and where it went. With experience we can tell how long ago it passed by and even whether it was injured. We are probably the only animal that can do this, and it is not far removed from the use of symbols to relate stories and information. Narrative is an important component in hunting, so we became the story-telling ape that looks for narrative and as an explanation of the past and a predictor of the future, and so was born the origin myths and legends that form the basis of collective cultures and group cohesion in common heritage, and the social ethics and group norms that are the glue holding society together.

The long childhood apprenticeship in these cultural norms also gave rise to another byproduct of childhood – childhood gullibility and the tendency to play safe and assume agency where there is none in preference to assuming nothing is causing a bush to shake, only to discover that it is being shaken by a python or a leopard.

Those who assumed agency and didn't take the sceptical approach and fact check, lived to tell the tale and pass their teleological thinking genes on to their children and those who accepted what their parents and authority figures in their group said as gospel truths, lived to breed because they didn't go to see if there really was a crocodile in the waterhole.

The argument that the ability of humans to speak in complex sentences to communicate idea is something uniquely human, which therefore must have been intelligently designed, is both a *non sequitur* and an argument from ignorant incredulity

The fact that languages evolved is demonstrated in the fact that there are so many of them, although to a small town in Kansas it might seem that there are only two -English and Spanish - in fact over most of Europe. Africa and Asia and in the Americas before the European conquests, there are or were, hundreds of languages and dialects which can be grouped, like species, into families and superfamilies. To take the major Eurasian languages alone:

Indo-European Languages: These are believed to have originated in the Pontic-Caspian Steppe (modern-day Ukraine and Russia) around 6,000–8,000 years ago. This is supported by the "Steppe Hypothesis," which links the spread of Indo-European languages to the expansion of Yamnaya pastoralists. (4) (5). The language family they brought with them has since diversified into:

> Germanic: English, German, Dutch, Scandinavian languages.
>
> Romance: Spanish, French, Italian, Portuguese.
>
> Slavic: Russian, Polish, Czech, Serbian.

Introduction

    Indo-Iranian: Hindi, Persian, Pashto.

    Celtic: Irish, Scottish Gaelic, Welsh.

    Hellenic: Greek.

    Baltic: Latvian, Lithuanian.

    Albanian and Armenian are standalone branches.

Uralic Languages: These possibly arose near the Ural Mountains about 6,000–8,000 years ago. They have sine diversified into:

    Finnish, Estonian, Hungarian, and minority languages in Russia (e.g., Komi, Mari).

Altaic Hypothesis. This contentious hypothesis suggests a grouping of Turkic, Mongolic, and Tungusic languages, and possibly Korean and Japanese. This is controversial, and these languages are often treated as separate families. Examples include:

    Turkic: Turkish, Uzbek, Kazakh.

    Mongolic: Mongolian.

    Tungusic: Manchu.

Sino-Tibetan Languages: These languages have their likely origins in the Yellow River region of China. They include:

    Sinitic branch: Mandarin, Cantonese, other Chinese dialects.

    Tibetan, Burmese, and numerous smaller languages.

Other Eurasian Language Families include:

    Dravidian: Tamil, Telugu, Kannada, and Malayalam, originating in southern India.

> Caucasian: Several unrelated families in the Caucasus region, such as Kartvelian (Georgian), Northeast Caucasian, and Northwest Caucasian.
>
> Isolates: Basque in Spain/France, whose origins remain unknown but predate Indo-European languages.

It is obvious from perusing that list of Eurasian languages that they fit more logically into the nested hierarchies that are seen in animal and plant species, forming distinct families showing an evolutionary relationship.

There is also evidence of hybridization and significant ingressions from other languages. For example, modern English was basically a Germanic language related to Dutch and Danish but received a large import of Norman French after 1066 when French became the Official language of the aristocracy, and the common people spoke English. There was also a strong Danish influence in the north of England under the earlier invasions of Danish Vikings and Danish rule in Northumbria.

Of course, the creationist explanation for different languages is to fall back on the implausible myth of the Tower of Babel and the alleged .confounding of the tongues of everyone who is supposed to be descended from that population just a few generations after the alleged biblical flood.

The nonsense of that story can be seen in the fact that the supposed designer god forgot that people can learn two or more languages and the fact that language distribution is based on linguistic families, not entirely unintelligible languages co-existing as neighbours as the tale suggests.

In reality, despite the plethora of Eurasian languages as listed above, it is possible to travel from Beijing to Madrid, during which you will pass through dozens, if not hundreds of different

Introduction

language areas, but nowhere will you find a place where a man cannot understand his neighbour.

Languages are not evidence of intelligent design. Looked at objectively, they are evidence of evolved human cultures and migrations.

Rejection of Evolutionary "Explanations for Flaws"

Evolution is rejected by creationists by dogma, not because there is objective evidence falsifying it, so any imperfections in the 'design' of human is simply waved aside and dismissed as misunderstandings or a failure to understand the motives of the designer, which are unknowable to us.

Which is another way of saying that creationism provides reasons to be satisfied with not knowing.

For example, vestigial organs are not regarded as evidence of evolved redundancy but are assumed to have some, as yet undiscovered, function.

Disease such as genetic and degenerative disorders, cancers or the failure of our immune system to defend us from attack by parasites are blamed, not on design faults and failures but on 'sin' following the 'Fall' in Christian theology.

A notable attempt to dress this up as real science was made recently by Discovery Institute fellow, Professor Michael J. Behe, who took the notion of 'genetic entropy' as proposed by Samuel A Cushman (2015) (6) and claimed this caused 'devolution' from some assumed initial perfection (7). The notion of devolution is biologically nonsensical because there is no mechanism for a deleterious mutation to increase in the gene pool of a species, and any mutation which makes a parasite better at parasitising its host can't logically be described as devolution. It is, in fact, classic evolution.

But Behe, is writing for an audience that has a parody version of what scientists believe is evolution, i.e. where individuals turn into other species, so imagining that evolution happens as a single event in one individual is well within their level of misunderstanding. That Behe and his colleagues in the Discovery Institute are attacking an intelligently designed straw man version of Darwinian evolution never occurs to their readers. Their objective, as defined in the Wedge Strategy (8) (9), is to destroy the American public's trust in science, using evolution as the thin end of the wedge to undermine the authority of science.

The purpose of this book then is to show that far from being the perfect design of a perfect, omniscient, omnipotent god, the human body is the result of 3.5 billion years of evolution with all that processes compromises, and utilitarianism where near enough is good enough and layers of complexity have been added to mitigate earlier bad designs to either compensate for when they fail or to improve an inefficient process that any intelligent designer would have scrapped as soon as its inefficiencies became apparent if it had lacked the necessary foresight to have designed it properly in for first place.

Looked at critically, beneath the superficial resemblance of design, the human body is not an argument for intelligent design but very much an argument against it.

For instance, genetic disorders are caused by a failure of the DNA replication mechanism which allow mutations to go uncorrected. DNA replication is an error-prone process for which correction mechanisms have evolved as an added layer of complexity which would not have been needed if the replication

Introduction

machinery had been designed by a prefect designer with the foresight to anticipate and allow for errors.

Yet the 'near-enough-is-good-enough' error correction mechanism is itself flawed. It works well enough most of the time and when it fails, it's often when the person is past reproductive age anyway or in earlier time before modern medicine, died in early infancy so the parents could produce another baby fairly quicky, so there was little evolutionary pressure to evolve anything better.

The skeletal compromises to allow bipedalism to develop which I mentioned earlier now give rise to a number of problems ranging from impacted wisdom teeth to complications in childbirth and slipped discs and debilitating chronic back pain, particularly in the elderly. The benefits of bipedalism so far outweighed the disadvantages of these skeletal compromises that the suboptimal solution was selected for by natural selection in preference to any deleterious effects of the compromises.

I will expand on these in a later section.

In my book, *The Unintelligent Designer: Refuting the Intelligent Design Hoax* (10) I defined good design as having:

- Clear functionality, so that the purpose of the design is clear to see and meets the requirements.
- Maximal simplicity and minimal complexity.
- Minimally wasteful in terms of energy and resources.

Bad design then is the antithesis of good design, vis

- Unnecessarily complex.
- Wasteful of energy and resources.
- Lacking a clear purpose and a failure to perform.

The Body of Evidence

In the following pages I will show how the human body, far from being the perfect design creationists like to imagine, is in fact the result of compromise and suboptimal solutions created by a mindless, 'make-do-and mend', 'near-enough-is-good-enough' natural process that has neither direction nor ultimate purpose that exists simply because it is good at producing the next generation, the purpose of which is to produce the next generation and so on – in fact the same reason every other living species exists, as the end-point of a natural evolutionary process that began when the first self-replicating molecule made the first copy of itself, and then there were two…

Introduction

Part I: Flaws in Basic Architecture

The Body of Evidence

Part I: Flaws in Basic Architecture

From the Top Down

In this section, I'll be taking a top-down approach, literally, by starting at the head and working down the skeleton to the feet. As I've mentioned before, most of the problems with our skeleton are the result of compromises made during the evolution of bipedalism when, as apes abandoned by the trees in what was quickly becoming the African savannah, we had to adapt to a new lifestyle, new food resources and perhaps most importantly, a new need to cover large distances as efficiently and quickly as possible,

Such were the advantages in being able to do these things under the circumstances that the resulting downside palled into insignificance in comparison. In economic terms, the cost-benefit ratio was strongly in favour of the benefits.

So, what were those costs?

The face and skull.

The face and skull are the most recent structures to have evolved to give us full bipedalism, a large brain and smaller teeth, having evolved together in the last few hundred thousand years. Evolution to bipedalism and an upright posture were more or less complete by 3.6 million years ago, judging by the Laetoli footprints (11) - still relatively recent in evolutionary terms but far longer ago than the remodelling of the skull, which anatomically distinguishes us most from our immediate ancestors and cousin species such as Neanderthals.

Looked at objectively, no matter how beautiful we perceive the human face to be, it is surely the strangest face in the mammal kingdom, with its flat face, high forehead and funny little triangular nose sticking out of the middle of it. As the brain enlarged and the foramen magnum[6] moved underneath to a central position to bring the skull over the spinal column instead of attached to the end of the horizontal spinal column in the typical mammal, so the bones of the face migrated down to bring the eyes down to look forward. Now, instead of sticking out of the front of the skull as the face does in say a dog or a cat, in the simians and apes, the facial bones now hang down beneath cranium – a trend which started in the simians, became extreme in *H. sapiens*.

The final stage in this transformation was a reduction in the size of the teeth as our diet became softer so needling less chewing when we discovered fire and started cooking.

Look at a typical human face profile and draw an imaginary line down the nose, following the curve round to the chin, and you will have drawn the profile of a typical monkey. Now imagine the muzzle pushed back to more or less in line with the vertical forehead and eye-sockets and you will have the human facial profile with its strange chin left protruding at the end of the lower jaw.

The result of this extensive remodelling of the face is to compromise drainage tubes leading from various sinuses to the nasal cavity. The function of these paired sinuses is debated but one of them is probably to lighten the bones they are embedded in. The linings of these sinuses produce a continuous stream of mucus that drains through tubes into the nasal cavity, the purpose of which is to wash away any bacteria that find their way into the sinus. If, as frequently happens during a viral infection the lining of the tubes becomes swollen, closing the tubes, so these sinuses can become blocked. This can set up

---

[6] The hole in the base of the skull through which the brain stem passes to become the spinal cord.

Part I: Flaws in Basic Architecture

ideal conditions for opportunist, anaerobic[7] bacteria to proliferate leading to sinus infections. These infections, thankfully rarely, can track along the trigeminal nerve to the brain where they can cause a potentially fatal meningitis.

If these sinuses become chronically infected, as they often do in childhood, they can cause the adenoid lymph glands to become swollen making it difficult to breathe through the nose and will result in a perpetual rhinitis and runny nose. Unless corrected surgically by removal of the adenoid tonsils, this can result in a distorted development of the facial bones to give the flattened face with flaring nostrils of the typical 'adenoid face' (12).

One of the tubes which can also become blocked is the Eustachian tube leading to the middle ear – the otherwise closed chamber between the ear drum and the inner ear where the auditory and balance organs reside. If this chamber can't communicate with the nasopharynx the pressure inside can't equilibrate with the pressure on the outer surface of the ear drum, causing the ear drum to bulge painfully. Pinch your nose and swallow to feel the effects of this disequilibrium.

The problem is that during a virus infection such as the common cold or a hyperallergic reaction such as hay fever (of which more later) sneezing can force mucus into the Eustachian tube along with opportunist bacteria so the middle ear become chronically infected and painful, resulting in very painful *otitis media* and 'glue ear' requiring the surgical creation of an alternative drainage tube with grommets in the ear drums. In rare cases these infections can lead to a mastoid abscess which can result in deafness.

---

[7] Able to live in low or no oxygen environments.

Unless the designer of this facial architecture intended us to get these painful and possibly even life-threatening infections, these are major design flaws.

Wisdom teeth

Lower down in the face, we have problems caused by the reduction in the size of the maxilla and mandible which has not been matched by the number and type of teeth. This can lead to problems with the last teeth to erupt – the molars or wisdom teeth for which there is often too little room in the gum, so the teeth come through at an angle or don't fully erupt.

I laughed out loud when a well know Young Earth creationistgrifter who appears to have no understanding of even basic science, Ray Comfort, complained to his followers in the social media about the pain he was suffering from an impacted wisdom tooth for which he needed surgery (13) He obviously had no understanding that he was giving his followers a wonderful example of the result of utilitarian evolution and something that couldn't possibly be attributed to an intelligent designer, unless that designer is also malevolent and is actively working to make people suffer.

Teeth are also notorious for being the breeding ground for the bacteria, *Streptococcus mutans*, that cause dental carries that, if left untreated can tunnel through to the dental pulp allowing access to other bacteria that can set up a painful abscess in the jaw or maxilla under the tooth. Other bacteria can get between the tooth and the gum where they consume the suspensory ligaments that hold the teeth in place, allowing them to fall out, making chewing more difficult. Some of these oral bacteria are highly organised colonial associations of different species that form the dental plaque (14), some of them are implicated in rheumatoid arthritis (15) and stomach cancer (16).

Part I: Flaws in Basic Architecture

Choking hazard

Moving further back in the mouth we have an example of how evolution gets stuck with a suboptimal design which would be impossible for Darwinian evolution to undo and start again with a better design. We find the 'fatal error', in the way the food pipe crosses the windpipe. In 2022, 5,554 Americans died from choking when a lump of food became lodged in their airway. In the same years 276 people died from choking in England and Wales (17). The reason this is a special problem in humans is because the top of the larynx is lower, relative to other mammals, probably as part of the evolution of speech, so it needs to move further to be closed by the epiglottis during swallowing.

There is of course, no intelligent design reason for this arrangement; it is, instead an example of suboptimal compromises and exaptation of a structure designed for a different purpose, as mammals evolved from marine vertebrates:

Ancestral origins in fishes: in fish, the pharynx served as a passage for both water (for gills) and food. In these aquatic ancestors, water entered the mouth and passed through the pharynx, where gills extracted oxygen. The same space was also used to transport food. Terrestrial tetrapods, including mammals, inherited this dual-purpose structure.

Transition to land: during the evolution of tetrapods, the breathing system adapted from gills to lungs, probably developing from the flotation chambers used as oxygen reserves by bottom-feeding fish that foraged in low-oxygen sediment, that would keep them refreshed by gulping air at the surface. The trachea evolved as a dedicated airway to the lungs, while the oesophagus continued to serve as a food passage. Both structures retained their ancestral connection through the pharynx. The redundant gill arches were then exapted for other purposes, one of which was the formation os the larynx

Speech and upright posture in mammals: in mammals, especially humans, there were further modifications: the larynx, which evolved from gill structures, descended lower in the throat to enable complex vocalizations, a key adaptation for speech. This descent of the larynx increased the risk of choking because it elongated the shared space between the airway and the food path. Our upright posture also affects this anatomy, compressing and rearranging the throat structures, contributing to the crossing paths.

This is a classic example of how, unlike an intelligent designer who is not restricted to modifying existing structures, evolution works by modifying existing structures rather than designing them from scratch. The shared pharynx is a "jury-rigged" solution resulting from our evolutionary history. It has persisted because, despite the risks (e.g., choking), the benefits of vocal communication, bipedalism and other adaptations outweighed the disadvantages in survival and reproduction. In other words, a utilitarian and suboptimal 'design' through an unplanned and purposeless natural process.

The cervical vertebrae

The next problem we encounter on our journey down our evolved skeleton is the cervical curve towards the top of our vertebral column. This needs to facilitate a highly mobile head, a consequence of our bifocal, forward looking eyes, with their relatively narrow field of vision, that mean we have to turn our head a lot to look around and up and down. Over a lifetime this can result in wear and tear on the joints between the vertebrae. One potentially fatal consequence of this need for high mobility is a result of the arrangement of the top two vertebrae, the atlas and axis, on the top of which the head sits. The atlas is basically a ring of bone that supports the head and rotates around a peg or tooth, the *dens* protruding from the axis (it is actually derived from the vertebral body of the atlas in the embryo). In a whiplash injury when sudden deceleration or acceleration causes the head to jerk backwards and forwards, this peg can snap, and the broken end can sever the spinal cord in a fatal injury.

Additionally, a blow to the top of the head can push the skull down into the atlas causing it to split, meaning the head must be fixed in a neck brace while the bone heals.

The lumbar vertebrae

The next serious evolved problem on our journey down the skeleton is another of the curves I mentioned earlier. The lumbar vertebrae. Like the cervical curve, this was necessary to bring our trunk into an upright position balanced on our legs for a fully upright posture. The curve is achieved by making the intervertebral discs wedge-shaped rather than flat discs as they are in other mammals.

With the weight of our upper body bearing down on these wedges, and our tendency to bend forward to lift objects instead of bending our knees and keeping the back upright, so using our back muscles rather than thigh muscles in lifting, this can easily cause the discs to herniate to give the painful and debilitating 'slipped disc', which can press on the sciatic nerves as they emerge from the spinal cord, causing the radiating pain in the backs of our legs known as sciatica. Chronic back pain, 'lumbago', can become an increasing problem in old age.

A generalised problem of the spine, not restricted to any one set of vertebrae, but common in thoracic vertebrae is the problem of scoliosis, or curvature of the spine, caused by the vertebral bodies crushing under the weight of the, body above them being born by the spinal column, unlike in quadrupedal animals where the spine is not used for weight-bearing, other than the weight of the internal organs suspended below them and supported partly by abdominal muscles. Turning the human spine through 90 degrees, caused the spine to become a weight-bearing structure that it had never evolved to be.

The pelvis and pelvic floor

Perhaps the most important result of the evolutionary compromises involved in the human upright posture is the narrow, curved birth canal, through which the large-headed human baby must be pushed. Even with modern medicines, good hygiene and nutrition, the birth process in humans is more difficult and prone to serious risks to the health of both mother and child, and probably more painful (so far as we can measure the pain experienced by other mammals), than any other mammal.

Relative to the size of baby that has to pass through it, the human pelvis is narrow, even though it is wider than in most other mammals of comparable size. This is a compromise between the pelvis being narrow enough for efficient walking and running, and wide enough for childbirth. The suboptimal compromise is that women still have a hard time giving birth, but their walking and running are not as efficient as they could be.

Something which again is more of a problem for women than for men who have a narrower pelvic, that is a consequence of the shoddy compromise between narrowness and childbirth, with the added complication of the abdominal contents pressing down into the pelvic because of the upright body, is the relatively weak pelvic floor. This is exacerbated by childbirth and makes women susceptible to a prolapsed uterus and stress incontinence. Basically, the mammalian pelvic floor was not designed through evolution to be a floor to keep the contents of the abdominal cavity from falling out.

Part I: Flaws in Basic Architecture

External testes

Although men have narrower hips, they too have a problem caused by the upright gait and the tendency of the weight of the abdominal contents to be borne by the pelvic floor. It is also a consequence of a more general design fault shared by almost all male mammals – the need for the testes to be kept at a lower temperature than other body organs for them to produce sperm, so they are carried externally in a pouch of skin.

To get there they descend normally soon after birth through holes in the pelvic floor together with the necessary blood vessels and tubes to convey the sperm to the epididymis where it is stored until ejaculated. This 'inguinal canal' is a weak spot in the pelvic floor.

This makes men vulnerable to an inguinal hernia where a loop of the small intestine descends through the inguinal canal into the scrotum. These require surgical correction as the intestine can become strangulated by the abdominal wall, leading to life-threatening complications.

External testes are also vulnerable to injury and if their temperature is raised by, for example tight clothing, this can result in a low sperm count and functional sterility.

The hip joints

Moving on now to the hip joint which has evolved to allow the legs to be in line with th spine instead, as with most vertebrates, at right angles to it. This has meant evolving a shallow socket or acetabulum for the major ball and socket joint of the body, to give the joint sufficient mobility. It has the additional stress of bearing the entire weight of the trunk, unlike in quadrupeds where the weight is distributed between front and back legs.

Consequently, it is relatively easy to dislocate and is subject to wear and tear resulting in chronic arthritic conditions. The top of the femur also has a prominent trochanter for muscle attachment which often bears the full weight of a sideways fall, resulting, particularly in elderly people who may have osteoporosis, in a broken neck of femur.

In earlier time before the advent of antiseptic surgery and anaesthetics, a broken neck of femur or a dislocated hip would almost certainly have proved fatal, if only because the victim would be quickly predated on by opportunist carnivores, but, because the victims are normally past child-bearing age, there was little evolutionary pressure to improve on this 'near-enough-is-good-enough' suboptimal compromise.

The knees

The knee joint is the largest joint of the human body and it needs to be since it bears the full weight of the body while performing twists and turns while partly flexed. In order to give it lateral stability, there are semi-lunar cartilages, or menisci, at each side and a pair of cruciate ligaments holding the femur and the top of the tibia together. These ligaments and cartilages are commonly damaged making walking difficult and running impossible. Again, in earlier time on the African Savannah, these might well have proved fatal unless the culture cared for the sick and injured which would have meant supplying them with food while they recovered, implying a surplus from hunting and gathering trips.

The ankles and feet

And finally, we come to the ankle and the foot which were part of our evolution to bipedalism. The ankle is a hinge joint that needs to bear the weight of the entire body and absorb the shock of landing on one foot during jumping and running and to

Part I: Flaws in Basic Architecture

transmit the propulsive power from the foot up into the leg when running and walking.

It is formed by the talus fitting into a more or less square shaped socket formed by the bottom of the tibia extending down one side as the medial malleolus and the end of the tibia forming the other side of the socket as the lateral malleolus. It is held together by radial ligaments from the ends of the malleoli to the heel bone or calcaneus and the bones of the foot.

The problem is, that these ligaments can be torn to give the 'sprained ankle' if the ankle joint is forced to bend sideways by stepping on the edge of a pavement, for example, and in a more severed twist, one or other (normally the lateral) tip of a malleolus can break by literally being pulled away from the rest of the bone by its attachment ligaments (an avulsion fracture). Some argue that the avulsion fracture is actually a sort of fail-safe mechanism since bone, especially vascular bone like the malleolus, will heal more quickly than a broken ligament.

Lastly, the foot itself is subject to orthopaedic problems such as fallen arches and plantar fasciitis.

The Body of Evidence

Part II: Internal Weaknesses

The Body of Evidence

Part II: Internal Weaknesses

The Needlessly Complicated Digestive Tract

Technically, the digestive tract begins at the lips and ends at the anus. I have already mentioned how problems with wisdom teeth are the effects of the evolution of the face failing to proceed in lockstep with the evolution of smaller teeth following the discovery of fire and the invention of cooking. I have also described how humans are especially vulnerable to choking because the design flaw in crossing the food pathway with the airway, is exacerbated by the lengthening of the pharynx to evolve complex speech.

There are also a number of other design flaws in humans, some related to evolving an upright posture; some to changes in our diet over time and some to general design faults.

Hiatus hernia

A hiatus hernia is a commonplace problem with the digestive tract which can cause pain and discomfort and restrict the sort of food that can be comfortably eaten. It can also result in a Barrett's Oesophagus (18), a condition that carries an increased risk of developing oesophageal cancer.

It is caused by the stomach herniating through the weak area in the diaphragm where the oesophagus passes through it. The diaphragm is a muscular structure which separates the organs of the thoracic cavity (heart and lungs) from those of the abdominal cavity and, with the intercostal muscles, aids in breathing. Over evolutionary history of mammals, the primary function of respiration has been prioritised over the secondary function – providing a seal around the oesophagus.

In humans, this compromise became a problem when an upright posture changed the direction in which gravity pulls on the stomach. This has exacerbated the weakness associated with the oesophageal passage. This problem tends to develop with age as the diaphragm is subjected to more stress and connective tissues become weaker. Coughing and sneezing, heavy lifting, straining with constipation and, in women, pregnancy, can all increase the risk of a hiatus hernia.

Gastric ulcers

The stomach is basically a storage organ where the process of digestion continues, having begun in the mouth with digestive enzymes in saliva. In the stomach, this process continues using hydrochloric acid and other enzymes which break down the carbohydrates and proteins. The problem is, the stomach is also made of proteins and carbohydrates, so, if not protected, the digestive juice in the stomach can erode the stomach wall causing ulcers which can lead to sever blood loss, anaemia and death.

Normally the stomach lining is protected by a layer of mucous which is continually renewed as quickly as it is digested, a neutralising bicarbonate secretion and tight junctions between the epithelial cells, however this protection is not infallible and stress, certain medications and infections can cause these defences to fail. On of the most significant cause is the commensal bacterium, *Helicobacter pylori*.

H. Pylori has coevolved with humans over about 100 thousand years, the details of which are outside the scope of this book, other than to say different strains mad on to different African populations and the different population of Europe and Asia that resulted from the migration out of Africa and diversification into the different populations worldwide.

It probably provides some symbiotic benefits by acting as a 'gatekeeper' to prevent infections by more dangerous pathogens. In evolutionary terms, these benefits outweighed the disadvantages of gastric ulcers which tend to develop in later years and do not affect reproduction enough to cause significant evolutionary pressure to develop defences again it.

An intelligent designer could, of course, have provided those defences while preventing the damage *H. pylori* can do.

So, gastric ulcers can be fully explained in evolutionary terms and utterly refute any notion of the involvement of an omnibenevolent intelligent designer.

Gall stones and risk of pancreatitis

Gall stones are a case study in unintelligent design. They form in the gall bladder, a small storage organ in which bile is stored until it is needed by the digestive system as food passes out of the stomach into the duodenum to continue the digestive processes started in the mouth and continued in the stomach.

Bile is a mixture of salts (bile salts) and the breakdown products of the haem group in dead red blood cells (the iron being retained and recycled) and excess cholesterol. In the duodenum it acts as an emulsifier of fat, breaking it down into small droplets that can be more easily attacked by digestive enzymes.

Gall stones can develop in the gall bladder when an excess of cholesterol relative to bile salts and phospholipids means cholesterol can crystallise and form stones, which are too large to pass through the bile duct. Instead, they can block the outlet from the gall bladder causing severe pain as pressure continues to build up in the gall bladder. This can result in an obstructive jaundice where bile builds up in the blood causing a yellowing (jaundice) of the skin and the whites of the eyes.

The gall stones can also become infected causing an infection to spread via the bile duct and its common outlet into the duodenum with the pancreatic duct in the 'ampulla of Vater' to the pancreas causing acute pancreatitis.

Gall stones provide ample evidence or poor design and retained functions that are no longer needed. Firstly, there is no need for the gall bladder to exist at all, as humans can live a perfectly healthy life following its surgical removal. While it may have had a function in an early ancestor it has none in humans, instead it introduces risks which can lead to death., which could all have been avoided if the bile had flowed directly into the duodenum.

The high cholesterol in bile salts may be a left-over from when we had a relatively low-fat diet so, in a modern diet there is more fat to be processes, consequently, there is more cholesterol in bile. The composition of bile depends on several factors such as diet, hormones and genetic factors. It would not have been beyond the wit of an intelligent designer to devise a more robust, stable regulatory system which would have avoided crystallisation of cholesterol altogether.

What has happened is that the slow evolutionary process has not kept up with changes in our lifestyle from one in which food was a variable quantity subject to periods of scarcity and periods of plenty, to one, at least in developed economies, where food is easily available and higher in fats than the diet of hunter-gatherers.

So, the problem of gall stones is readily understandable as the result of an evolutionary process with it compromises and trade-offs, especially when related to the history of our species. It is incomprehensible in terms of intelligent design.

Diabetes.

Like gall stones, diabetes is another case study supporting evolution over intelligent design. It occurs in two forms:

Type I – An autoimmune condition in which the body's immune system attacks the insulin-producing cells in the islets of Langerhans in the pancreas, causing a lack of insulin.

Part II: Internal Weaknesses

Type II – A metabolic disorder where cells become resistant to insulin, often accompanied by insufficient insulin production, linked to obesity, inactivity, and genetic predisposition.

Both of these result in hyperglycaemia which if uncontrolled leads to heart disease, kidney damage, nerve damage, atherosclerosis and blindness.

Diabetes highlights a number of flaws in human physiology of which an intelligent designer would have been aware and could have avoided:

Inadequate blood sugar regulation: blood sugar level needs to be maintained within a narrow range yet the system for doing so is prone to failure. The two hormones, insulin and its counter-regulatory hormone, glucagon, need to function precisely because even a minor imbalance can cause the blood sugar level to stray outside the narrow range.

For a life-critical system, an intelligent designer could have provided a more robust, fail-safe system or a compensatory mechanism for when it failed.

Autoimmunity in Type 1 diabetes: Type I diabetes is the direct result of a failure of the immune system, as are many other auto-immune conditions. This reflects a poorly 'designed, over-sensitive immune system that can and does make mistakes. It is a blunt instrument lacking precision and yet still overly complicated and error prone – not the work of an intelligence with foresight or concern for the efficiency of its designs. A designer indistinguishable from a mindless, near-enough-is-good-enough, make-do-and-mend natural process in fact.

Type II diabetes and an evolutionary legacy: Type 2 diabetes often results from the body's inability to cope with modern diets high in calories and processed carbohydrates. This mismatch between our evolutionary history and current lifestyles demonstrates how evolution adapts organisms to specific environments rather than optimizing them for all conditions. What suited our ancestors on the African savannah is no longer appropriate for life in a mechanised, developed economy.

Excess fat impairs insulin sensitivity, making Type 2 diabetes closely linked to obesity. Evolution for a hunter-gatherer lifestyle favoured fat storage in times of food surplus, as a survival mechanism in times of food scarcity, but in modern environments with constant food availability, this adaptation becomes a liability.

For hunter-gatherer groups, food supplies would have been unreliable, so during periods of famine, fat people would have been the last to die, to repopulate the gene pool with their fat-storing genes. Not surprisingly then, we have an evolved tendency to over-eat and store the surplus as fat; but for most people, unless they diet consciously, the lean times never come. The result, in advanced economies is a strong tendency toward obesity and so to Type II diabetes.

The process of evolution has no ability to plan for a different range of scenarios – the hallmark of an intelligent designer – so it can and does result in a system or process which is beneficial in one scenario but detrimental in another. Using insulin and glucagon to regulate blood sugar is an example of how evolution can only make use of what it has and doesn't invent and design anew, as an intelligent designer would, so we have ended up with a suboptimal, flawed blood-sugar regulatory process rather than the flawless, purpose-built process an intelligent designer would have provided.

If an intelligent designer is not designing for the future, then it can't be described as intelligent and calling it a designer is stretching the definition beyond breaking point.

Diabetes, with its roots in evolutionary trade-offs and its prevalence in modern societies, starkly demonstrates the imperfections of the human body. Rather than supporting the notion of a perfectly designed system, the disease highlights how evolution produces workable but flawed solutions. This makes diabetes a compelling argument against intelligent design and a vivid illustration of evolution's "good enough" engineering.

Paralytic ileus

Continuing with the problems a lack of intelligence in our design causes us, I'll turn now to the potentially life-threatening condition known as the paralytic ileus. In this condition, the normal peristalsis of the gut, or a section of it stops, leading to a build-up of part-digested food, gasses and liquids, abdominal bloating, pain, and complications such as bowel obstruction or perforation.

A paralytic ileus is typically caused by trauma, particularly abdominal trauma, surgery (usually but not always abdominal surgery), infections, for example peritonitis, medication, especially opioids and anticholinergics, and electrolyte imbalance.

While this condition highlights serious vulnerabilities in the design of the human digestive system, its evolutionary basis can be understood as an adaptation to acute stress or injury. However, like many biological responses, it reflects evolutionary trade-offs rather than an optimal design.

It's probable that the paralytic ileus response evolved in a remote mammalian or even pre-mammalian ancestor when shutting down the digestive tracts in times of injury or stress was beneficial in that it increased the chances of survival and recovery to continue to breed. It might have conveyed a number of benefits such as:

Protection against further damage: By reducing leakage lower down in the gut, the resulting peritonitis could have been reduced in extent and severity.

Conservation of energy: by shutting down the digestion processes during shock, blood and oxygen are diverted to essential organs such as muscles, brain, heart and kidneys aiding short-term responses.

Inflammation-Driven Response: Inflammation from injury or infection triggers a cascade of chemical signals, including cytokines and prostaglandins, which inhibit peristalsis. This again could be an evolved mechanism for shutting down a non-essential (at least in the short term) process to divert resources to a deal with a short-term threat.

Part of the Stress Response: Paralytic ileus is linked to the gut-brain axis, where the autonomic nervous system regulates gut activity via the vagus nerve[8]. The enteric nervous system, part of the autonomic nervous system, is highly sensitive to stress signals, such as those released after trauma. It is the system responsible for the 'fight or flight' response where a series of automatic nerve responses and a squirt of adrenaline and noradrenaline from the adrenal glands puts the body into emergency action mode.

While paralytic ileus may have provided some evolutionary benefits, it also exposes significant flaws in human anatomy and physiology. In particular it exposes how the modern human body is over-sensitivity to disruption. The digestive system is controlled by a complex of coordinated nervous, hormonal and muscle activity, the disturbance of which can cause an inappropriate and undesirable reaction that we haven't yet

---

[8] The vagus is the longest nerve of the autonomic nervous system in the human body and comprises both sensory and motor fibres. (Wikipedia)

Part II: Internal Weaknesses

evolved a better control of. Instead, we have a far from optimised system, based on what might have been beneficial to a remote ancestor in a different scenario but which was clearly not purpose-built for the needs of modern *Homo sapiens*.

A purpose-built system would, for example be able to distinguish context, instead ours does not differentiate between situations where halting digestion is beneficial (e.g., acute infection) and where it is harmful (e.g., postoperative recovery). A paralytic ileus is thus a maladaptive response in modern settings, such as after routine surgeries.

The autonomic nervous system is a primitive system for controlling basic life-support processes such as pulse rate, blood pressure, the automatic operation of internal organs such as peristalsis in the digestive tracts, but it lacks finesse and can easily be disrupted by stress, inflammation and chemical imbalance and the actions of higher centres in the central nervous system.

Paralytic ileus reflects the consequences and constraints imposed by earlier evolutionary compromised and suboptimal solutions. In particular, the digestive system shares neural and hormonal regulation with other systems. Evolution could not "start from scratch," resulting in a system that prioritizes survival during trauma but introduces vulnerabilities. Evolution favoured mechanisms that allowed quick suppression of digestion during stress or injury, even if it made the system prone to dysfunction in less dire circumstances.

And again, as with earlier problems such as diabetes, we see a mismatch between past needs and present needs, because of the inability to foresee and plan ahead - which would be characteristic of an intelligent design process - but which typifies a mindless, utilitarian evolutionary process. So, the paralytic ileus refutes intelligent design while demonstrating the suboptimal processes that the mindless process of natural evolution, devoid of intelligent input, produces.

This evidence against intelligent design can be summarised with:

> High failure rate: The digestive system's reliance on fine-tuned coordination between nerves and muscles makes it prone to paralysis, even from minor disruptions like electrolyte imbalances or medications.

> Lack of backup mechanisms: If peristalsis stops, there is no alternative mechanism to move contents through the intestines, leading to severe and potentially life-threatening complications.

> Harmful responses in non-lethal scenarios: The cessation of intestinal activity during surgery or minor trauma causes unnecessary suffering and complications, with no clear benefit.

> No long-term adaptive advantage: While paralytic ileus may have been useful in the distant past, its role in modern humans is almost entirely negative, reflecting the limitations of evolutionary adaptations.

Hence, a paralytic ileus is a prime example of how evolution produces "good enough" solutions rather than optimal designs.

While it may have evolved as a protective response to injury or inflammation, its persistence in modern humans highlights the flaws and trade-offs inherent in the evolutionary process. A truly intelligent design would likely include a more robust digestive system with fail-safes to prevent dysfunction, making paralytic ileus another argument against the notion of a perfectly engineered human body.

Part II: Internal Weaknesses

The appendix

Moving on down now, I need to give a passing nod to the appendix, that iconic atavistic fossil of herbivorous species somewhere in our remote ancestry when our digestion needed the assistance of microorganisms to digest the cellulose in a vegetarian diet.

It is attached at the bottom of the caecum close to its junction with the large intestine and is now nothing more than a wormlike blind sack (the vermiform appendix) which in the days before antiseptic surgery, could sometime become infected and burst, flooding our peritoneum with bacteria and digestive enzymes to kill us in a matter of days. It's hard to imagine now that acute appendicitis was frequently a terminal illness (19).

Although it is rich in lymphatic tissues which may play some part in our immune system, but which is more likely a feeble attempt to subdue a potentially fatal infection or an example of exaptation, and although there is some indication that the appendix may act as a repository for our microbiome, we can live perfectly healthy lives without it.

In fact, so unpredictable is its capacity to kill us that anyone intending to spend any length of time remote from civilisation is advised to have it removed prophylactically to abolish any risk of it doing so.

There is of course no sane reason why an intelligent designer would include a useless structure capable of killing us for no reason other than to include that danger in our design, but the appendix is not the only structure that can be understood as the product of mindless evolution and not at all as the design of an omnipotent, benevolent intelligent designer.

So, let us now pass on into the large intestine where yet more design errors await.

Ulcerative colitis, etc.

Ulcerative colitis (UC) is a chronic inflammatory condition affecting the colon and rectum, characterized by periods of remission and flare-ups. It is part of the spectrum of inflammatory bowel diseases (IBD), along with Crohn's disease. UC involves persistent inflammation, ulceration, and damage to the colon's inner lining, leading to symptoms like abdominal pain, diarrhoea (often bloody), weight loss, and fatigue.

Examining ulcerative colitis through the lens of the evolution vs. intelligent design debate reveals how the condition exposes vulnerabilities and inefficiencies in human biology. These flaws suggest a system shaped by evolutionary trade-offs rather than a perfectly designed system.

The key features of ulcerative colitis are:

> Inflammation limited to the colon and rectum.
>
> Affects the mucosal (innermost) layer of the gut.
>
> Symptoms include bloody diarrhoea, abdominal pain, and urgency to defecate.

The precise cause is unclear but is believed to be the result of an overactive immune response triggered by genetic predisposition and environmental factors, such as diet, stress, or infections. It produces an increased risk of colon cancer and complications can include a toxic megacolon (life-threatening dilation of the colon) and malabsorption, dehydration, and systemic inflammation.

Ulcerative colitis is probably the result of evolutionary trade-offs and constraints in the immune and digestive systems. It is an example of how adaptations that were advantageous in ancestral environments can become liabilities in modern ones because, unlike an intelligent designer, the evolutionary process can't plan

Part II: Internal Weaknesses

for the future and only responds to the here and now with what there is available to work on.

The immune system evolved to defend against pathogens, but in the case of UC, this defence becomes counterproductive. The immune system attacks the gut's lining, mistaking normal intestinal bacteria or tissue for harmful invaders. I'll say more about the immune system and the absurdity of a designer having an arms race with itself, in a later section. This was the subject of my last book, *Unintelligently Designed Arms Races: How Nature Refutes Intelligent Design* (20)

This represents a significant trade-off in that a robust immune system that protects against infections can also increase the risk of autoimmune diseases and inflammatory disorders.

There is a possible explanation for why this system should now produce an over-reaction resulting in ulcerative colitis in improvements in hygiene. Modern cleanliness reduces early exposure to microbes, leading to poorly regulated immune responses. In other words, the immune system is denied the opportunity to refine its response. In the ancestors in which the immune system evolved, frequent exposure to pathogens and parasites would have been the norm and would have helped the immune system develop tolerance.

A system which evolved for the unhygienic conditions we might well have been exposed to in childhood, way back in our remote pre-hominin ancestry is no longer fit for purpose in the environment those of us who live in developed economies are now brought up in, with a different microbiome due to different diet and lifestyle and a possible breakdown in the co-evolved symbiotic relationship we once had with our commensal gut flora and fauna. Unlike an intelligent designer, the evolutionary process had no means of anticipating these future changes and planning for them.

The human gut evolved under the constraints imposed by its earlier forms in ancestral vertebrates. As a result, it relies heavily on a delicate balance between microbes, immune tolerance, and structural integrity. This interdependence makes the gut highly vulnerable to disruption, as seen in UC.

The resulting flaws in the design of this system are something an intelligent designer could have avoided by building in a process to prevent the immune system attacking the body's own tissues and the immune system's failure to differentiate between harmful and harmless stimuli is a significant flaw of which any intelligent designer would be ashamed.

But why is the gut dependent on the microbiome in the first place? Almost certainly because what began as parasites in the digestive tract of the earliest creatures with a gut, evolved to become symbiotic by providing something in return for food and shelter and they have since co-evolved to become integral to the process of digestion and defence against other parasites.

But an intelligently designed process could have avoided the down-side of this Heath-Robinson[9] contraption by not being so dependent on exposure to potential parasites to 'train' it in the first place. The second blunder is the lack of a repair mechanism. Even if it couldn't avoid designing a system that was prone to malfunction, an intelligent designer could have designed a back-up repair mechanism. Instead, the body's evolved repair system is insufficient to prevent long-term complications like cancer.

---

[9] Named after William Heath-Robinson, an English illustrator who drew ridiculously over-complicated solutions to simple, everyday problems, often utilising household objects intended for a different purpose, in a manner highly reminiscent of how evolution exapts structures evolved for a different function.

Part II: Internal Weaknesses

To make matters worse, the colon is arguably a structure which is becoming redundant. Its purpose is to reabsorb water and recover the electrolytes poured into the gut with the digestive juices, and in humans, to store the faeces until they can be conveniently expelled, but it is not critical for survival. In severe cases of ulcerative colitis, it can be surgically removed and replaced with an ileostomy or J-pouch.

This raises the obvious question of why any intelligent designer would design such a complex and failure-prone organ. As the design of a putative intelligent designer, all ulcerative colitis does is causes chronic pain, disability, and a significantly reduced quality of life. An intelligent designer would likely create a system that minimizes such suffering.

Intelligent design advocates need to explain in what sense of the words 'intelligent' and 'design' can something be described as intelligent design when:

> It is so poorly adapted to modern lifestyles, that there is an increase in UC cases due to environmental and dietary changes

> It is failure-prone because the interplay between the immune system, gut lining, and microbiome is so complex that minor disruptions can lead to major disease.

> It results in an increased cancer risk:

By contrast to the inexplicable results of a hypothetical intelligent design explanation, the natural process of evolution explains it perfectly:

> Evolution favours traits that improve reproductive success in specific environments, even if they lead to vulnerabilities later in life. The immune system's overactivity in UC reflects this trade-off.

The rapid shift from ancestral environments to modern lifestyles has exposed the limitations of evolutionary adaptations. The rise of UC in industrialized societies highlights the consequences of this mismatch.

The gut's anatomy and function are shaped by evolutionary constraints, such as its dependence on microbial interactions and its inability to regenerate effectively after chronic damage.

Like the earlier examples of gastric ulcers, gall stones, diabetes and the paralytic ileus, ulcerative colitis exemplifies the flawed nature of the human digestive and immune systems. Rather than reflecting the work of an intelligent designer, the disease highlights evolutionary compromises and trade-offs that leave humans vulnerable to chronic inflammation and its complications.

The condition underscores how evolution produces systems that are "good enough" for survival but riddled with inefficiencies, making it a strong argument against intelligent design.

Irritable Bowel Syndrome

Irritable bowel syndrome (IBS) is another example of how the human body reflects evolutionary compromises rather than intelligent design. While IBS is not a structural or immune-mediated condition like ulcerative colitis, it is a chronic functional disorder that underscores inefficiencies in the design of the human digestive system.

It is a common disorder affecting the large intestine, characterized by symptoms such as:

Abdominal pain or cramping.

Changes in bowel habits (Diarrhoea, constipation, or both).

Part II: Internal Weaknesses

Bloating, gas, and discomfort.

Mucus in the stool.

IBS does not cause structural damage or inflammation like ulcerative colitis but results from dysfunctional interactions between the brain, gut, and microbiome.

IBS is an embarrassment for intelligent design advocates because it is caused by:

An oversensitive gut-brain axis, being closely linked to dysregulation of this axis, where signals between the nervous system and gut are mismanaged. This oversensitivity can lead to pain or discomfort from normal intestinal activity, indicating poor optimization in neural and muscular coordination.

It is an inappropriate stress response in which symptoms are often triggered or exacerbated by stress, suggesting an evolutionary trade-off where heightened stress responses—beneficial for survival in ancestral environments—have become maladaptive in modern, low-threat settings. An "intelligent" design could include a better stress modulation system to prevent non-lethal stressors from causing severe physiological disruption.

An excessive gut sensitivity, so that people with IBS often experience "visceral hypersensitivity," where normal gut stretching or activity feels painful. This exaggerated sensory response is an inefficiency that causes unnecessary suffering.

IBS is linked to imbalances in the gut microbiome (dysbiosis), which can arise from diet, antibiotic use, or other environmental factors. The heavy dependence of the gut on microbiota for normal function exposes a design flaw: the system is too easily disrupted by external changes and can be aggravated by diets low in fibre and high in processed foods, which evolution did not anticipate.

Unclear Triggers and Lack of Resolution: Unlike acute diseases with clear causes and cures, IBS is chronic and unpredictable. There is no single treatment, and managing symptoms often requires trial and error. This lack of reliability reflects evolutionary compromises rather than a well-planned, optimized system.

Like so much else about the human body, the gut evolved to handle a varied diet, respond to pathogens, and maintain homeostasis, over a very long evolutionary history, predating humans by hundreds of millions of years. And more recently on the African savannah for some six million years following our divergence from the last common ancestor we share with chimpanzees. However, in adapting to diverse and changing environments, it became prone to overreactions and miscommunications, as seen in IBS.

In particular, traits like stress-induced gut responses (e.g., "fight or flight" diarrhoea) likely provided survival advantages in dangerous situations but are maladaptive in modern sedentary lifestyles.

And as we've met before there is a mismatch between what was needed then and what is needed for a modern lifestyle. In particular, IBS is exacerbated by sedentary behaviour, processed diets, and high stress, but evolution has not had time to adapt to these modern pressures.

The consequence of the incremental evolutionary changes, using what was available and unable to create from scratch, is that the human gut-brain axis and microbiome-dependent digestion, is a system that is overly complex and failure-prone. An intelligent designer would not have depended on the efficient functioning of such a fragile, error-prone solution to a need.

What we see in IBS is a flawed system that causes chronic suffering without any benefit. It offers no evolutionary advantage, nor does it serve any beneficial function for the individual. It leads only to discomfort and reduced quality of life.

It is a system shaped by 'trial and error' so the symptoms vary widely between individuals, and their triggers can be hard to pinpoint. This inconsistency is a hallmark of a system shaped by evolution rather than purposeful design.

IBS lacks a cure, and treatments focus on symptom management rather than addressing the root cause. This reflects the body's inability to self-correct the dysfunction.

Digestive discomfort, diarrhoea, and malabsorption caused by IBS can result in wasted energy and resources, further undermining the notion of a well-designed system.

IBS is a classic example of what characterises bad design, as I spelled out in the Introduction: unnecessary waste, needless complexity and no clear purpose. Irritable bowel syndrome then provides another compelling argument against intelligent design.

Its chronic nature, susceptibility to stress and lifestyle factors, and lack of a clear biological advantage highlight the inefficiencies of the human digestive system. The condition is a vivid example of how evolution produces systems that are functional but flawed, subject to trade-offs and mismatches with modern environments. IBS exemplifies the evolutionary compromises that result in unnecessary suffering, undermining claims of perfect, purposeful design.

Diverticula and Diverticulitis

Before we get to the end of the digestive tract, literally, there is one more condition that takes phenomenal mental gymnastics to present as the result on intelligent design. This is the formation of diverticula in the wall of the large intestine and the diverticulitis that sometimes develops in them.

Diverticula are small, bulging pouches that can form in the lining of the digestive tract, most commonly in the sigmoid colon. When these pouches become inflamed or infected, the condition is called diverticulitis. Both diverticula and diverticulitis highlight significant flaws in the design of the human digestive system and provide strong arguments against the notion of intelligent design.

As, hopefully, you've come to expect by now, these are the result of a flawed, utilitarian evolutionary process full of suboptimal compromises and trade-offs and the antithesis of what we would expect of an intelligent design process.

Diverticula form in areas of weakness where blood vessels penetrate the muscular wall of the colon. An "intelligent" design would reinforce these vulnerable areas to prevent bulging under pressure. They are strongly associated with a low-fibre diet, which increase intraluminal pressure during bowel movements.

Evolution has optimized the colon for a high fibre ancestral diet, and of course lacked any mechanism for planning ahead for modern low-fibre diets, so this has now emerged as a design flaw. What worked in the East African savannah no longer works in New York, London, Paris, or Rome. An intelligent designer could have designed a colon that would tolerate dietary variability without developing diverticula.

A basic problem is the sigmoid colon's looping structure which creates regions prone to stagnation and high pressure, increasing the risk of diverticula. This redundancy in the colon's shape serves no clear purpose and introduces unnecessary vulnerabilities.

Once diverticula form, they are prone to trapping food particles and bacteria, leading to inflammation and infection. The absence of protective mechanisms in these pouches is a glaring oversight in any notion of intelligent design.

Diverticulitis can lead to:

> Perforation of the colon wall, causing life-threatening peritonitis.
>
> Fistulas, or abnormal connections between organs.
>
> Abscesses requiring surgical drainage.
>
> Bowel obstructions necessitating emergency surgery.

These complications serve no adaptive purpose and impose significant suffering, highlighting poor "design."

In contrast again to the incomprehensibility of any intelligent design explanation, the evolutionary explanation is readily understandable:

Early humans consumed diets high in fibre, which kept stools soft and reduced pressure on the colon. The evolution of a large, looping colon made sense in this context as it increases the surface area for reabsorption of water and electrolytes. However, a modern diet, low in fibre and high in processed foods, have exposed this system's vulnerabilities, showing how evolution adapts to past environments rather than optimizing for future conditions.

Diverticulosis and diverticulitis are more common in older adults, suggesting these weaknesses were less relevant in early human evolution when lifespans were shorter, and, since evolution prioritizes reproductive success over long-term health, this leads to systems that deteriorate with age.

And it's the same old story with evolution, where structures are often the vestige of suboptimal compromises, reflecting the random, incremental nature of evolution driven by the here and now environment rather than the result of intelligent design.

So, once again we see in diverticula no purposeful role having no known function in human physiology. Their formation and potential for inflammation represent purely detrimental conditions. Diverticulitis causes significant pain and can lead to life-threatening complications, all for no discernible benefit – the antithesis of intelligent design, unless the intention was to maximise the suffering in the world.

And there is that old, familiar mismatch between what was needed then and what is needed now, because, unlike an intelligent designer, evolution can't plan for the future.

The formation of diverticula and the condition of diverticulitis are clear examples of evolutionary inefficiencies rather than intelligent design. The colon's susceptibility to structural weaknesses, inflammation, and complications reflects its evolutionary history, shaped by trade-offs and adaptations to past environments.

These flaws highlight the haphazard, trial-and-error process of evolution, making diverticular disease another compelling argument against the notion of an omniscient designer and in favour of evolution as the explanation for the human body's imperfections.

Haemorrhoids

Lastly at the end of the digestive tract we have another consequence of an upright posture, piles or haemorrhoids.

Part II: Internal Weaknesses

One excuse made by creationists, for the nasty things like parasites, and all the results of defects we looked at so far, is that they are not the work of their designer god but the result of 'genetic entropy' causing 'devolution', so a word of caution for any creationists thinking of using that biologically non-sensical excuse, the Bible leaves no room for doubt that Yahweh creates 'emerods', as he inflicted them on the Philistines who stole the Ark of the Covenant from the Israelites, according to the tale in 1 Samuel 5:1-12.

Haemorrhoids, commonly known as piles, are swollen and inflamed veins in the rectum or anus. They can be internal (inside the rectum) or external (under the skin around the anus) and cause symptoms like pain, itching, bleeding, and discomfort.

These swollen veins are a direct result of structural weaknesses and vulnerabilities in the human body's design, particularly in the context of bipedalism, making them a strong argument against any notion of intelligently designed perfection.

Haemorrhoids are vascular cushions in the anal canal that help maintain continence and control stool passage. Under normal conditions, they are functional and healthy. However, when subjected to prolonged pressure or strain, they swell, become inflamed, and may prolapse or thrombose.

Prolapsed haemorrhoids can be caused by:

> Chronic straining during bowel movements (often due to constipation or low-fibre diets).

> Increased abdominal pressure (e.g., from pregnancy, obesity, or prolonged sitting).

> Aging, which weakens the connective tissue supporting the rectal veins.

Prolonged sitting or standing, which increases venous pressure due to gravity.

When human switched to upright, bipedal locomotion this drastically increased pressure on the veins in the lower body, including the rectum and anus. Unlike quadrupeds, where blood return from the lower body encounters less gravitational resistance, humans suffer from chronic strain on these veins, predisposing them to haemorrhoids.

The veins in the anal canal lack the strong structural support found in other parts of the body, making them more prone to dilation and swelling under pressure. This problem is exacerbated by aging, as connective tissues naturally degrade over time, leading to a higher incidence of haemorrhoids in older individuals.

Another problem is one we've met before in other parts of the digestive tract – the switch from the high-fibre diet for which our evolution as a hunter-gatherer had optimised us to a modern low-fibre diet. This means we are now prone to constipation, increasing the risk of haemorrhoids. In other words, it's the old mismatch between then and now because evolution can't plan ahead.

This would not have been a problem for an intelligent designer because intelligent design is design for future need, not for a now-redundant need. A clever design would have allowed for flexibility in diet, not be optimised for one particular diet.

Haemorrhoids confer no evolutionary or functional advantage. Their development only causes pain, discomfort, and in severe cases, significant medical complications like thrombosis or anaemia due to chronic bleeding.

Haemorrhoids have no adaptive function or benefit so can only be regarded as suboptimal design and structural defects. When

Part II: Internal Weaknesses

inflamed, haemorrhoids interfere with normal defecation and continence, causing significant discomfort and reducing quality of life. Any intelligence involved in their design can only be described as malevolent. As creationists can read in their Bible, their god allegedly created them for the purpose of making the Philistines' life a misery.

The fundamental design problem is that veins in the anal region have no efficient mechanism to counteract gravitational pull, leading to blood pooling, increased venous pressure, and eventual swelling. An intelligently designed system might include stronger venous valves or better drainage mechanisms to prevent hemorrhoidal swelling.

The transition to bipedalism was a major evolutionary step, allowing for efficient locomotion and freeing the hands for tool use. However, it came with trade-offs, including increased strain on the lower back, knees, and pelvic venous system. Haemorrhoids are a byproduct of this compromise, reflecting the imperfect nature of evolutionary adaptations.

Again we see that something that evolved to suit our ancestors has failed to keep up firstly with our rapid evolution of an upright, bipedal body and later with our rapid cultural evolution bringing a new lifestyle, new diet and different stresses of modern life, none of which we were 'designed', either by evolution or intelligence, for.

Just as with the other degenerative and wear-related defects that plague us. evolution prioritizes traits that enhance reproductive success and survival to reproductive age. Haemorrhoids often develop later in life, after reproductive years, when evolutionary pressures to optimize function diminish.

Haemorrhoids are another glaring example of poor design in the human body. Their high prevalence, painful symptoms, and lack of adaptive purpose underscore the inefficiencies of the human vascular and digestive systems. They are the result of evolutionary trade-offs, particularly the shift to bipedalism, rather than evidence of intelligent design. This condition vividly illustrates how evolution produces systems that are functional but flawed, leaving humans vulnerable to unnecessary suffering.

Part II: Internal Weaknesses

The Delicate Heart and Circulatory System

Having gone through the digestive tract from top to bottom, I'm now going to look at the examples of bad design to be found in the circulatory system. The circulatory system is the primary transport system of the vertebrate body. It distributes nutrients, water, and hormones, delivers oxygen to the cells, and collects waste such as carbon dioxide to be expelled via the lungs in exchange for fresh oxygen. It is also responsible for providing the platelets and clotting factors that plug holes caused by injury to maintain homeostasis.

Blood clotting and thromboses

While clotting is essential for survival, it is also prone to significant failures, such as excessive clotting (thrombosis) or insufficient clotting (bleeding disorders). The system's delicate balance and frequent malfunctions point to an evolutionary history full of trade-offs and inefficiencies rather than an intelligent design.

In the introduction, we saw why the claim by intelligent design advocates that the clotting cascade is an irreducibly complex process that proves intelligent design is merely an argument from ignorant incredulity and gap-filling with a presupposed designer god.

We'll see now why the creationist claim is a misrepresentation of what in reality is evidence of an evolutionary process including gene duplication and repurposing, excessing error-prone complexity and all the attendant suboptimal compromises and inability to plan for the future that we've seen in the evolved, unintelligently designed digestive system.

The coagulation cascade involves multiple pathways (intrinsic and extrinsic) and more than a dozen proteins. This redundancy likely arose through evolution as simpler systems were co-opted and expanded rather than being designed from scratch. An intelligently designed system could achieve clotting with fewer steps, reducing the chance of failure or miscommunication.

One of the potentially serious, life-threatening risks is that of the formation of thromboses. A thrombosis occurs when clots form unnecessarily, blocking blood flow. This can lead to life-threatening conditions such as:

> Deep vein thrombosis (DVT) where clots form in the deep veins, often in the legs.

> Pulmonary embolism (PE), where a clot travels to the lungs, blocking oxygen exchange.

> Stroke or myocardial infarction, where clots in arteries leading to the brain or heart cause catastrophic damage.

A better design would prevent clots from forming without an injury, while still responding quickly to actual damage.

The complexity of the clotting cascade means there are numerous points where genetic mutations can lead to disorders like haemophilia, where blood fails to clot adequately.

Such vulnerabilities suggest a poorly optimized system, as survival can depend on the integrity of every single step.

As humans age, their clotting systems become increasingly prone to malfunction, because:

> Arteries harden due to plaque buildup (atherosclerosis), increasing the risk of clots in the circulatory system.

> Sedentary lifestyles exacerbate venous blood pooling, increasing the risk of DVT.

These conditions are avoidable in an intelligently designed system that could adapt to aging and environmental changes.

Part II: Internal Weaknesses

Platelets, essential for clot formation, are prone to overreacting. Even minor damage to blood vessels or endothelial cells can trigger clotting, sometimes leading to complications like microvascular thrombosis, which can block capillaries.

As always, these problems can be explained as the result of evolution over the long ancestry of our species and as an injury-prone species eking out a precarious existence in the African savannah.

Evolution shaped the clotting system for survival in environments where injuries were frequent and fatal. In modern settings, where physical trauma is less common, the system's predisposition toward clotting can be more harmful than helpful.

The clotting system evolved in stages. Simpler organisms, like jellyfish, have rudimentary systems to seal injuries. In vertebrates, the system grew more complex as additional steps were added for precision, redundancy, and speed. These incremental changes resulted in a system that is functional but overly complex and prone to errors.

As we've seen before, evolution prioritizes survival over perfection. A system that clots "too easily" was more likely to save individuals from fatal bleeding after injury, even if it increased the risk of thrombosis later in life. Such trade-offs make sense in evolution but contradict claims of optimal, intelligent design.

And of course, there was the inability to anticipate future needs that characterises a mindless evolutionary process and gives the lie to claims of intelligent design. Humans evolved in environments where movement was constant, minimizing risks like venous stasis. Modern sedentary lifestyles expose the flaws in a system not adapted for prolonged inactivity, leading to conditions like DVT and pulmonary embolism.

An intelligently designed system would suffer from none of these flaws. It would be a simpler, more robust clotting system which would achieve the same results without involving so many components, reducing points of failure. For example, rather than cascading activations of clotting factors, a single, direct response mechanism could suffice.

An intelligent design would include mechanisms to better distinguish between injury and non-threatening conditions, avoiding unnecessary clotting (e.g., thrombosis) and would include stronger safeguards against runaway clotting, such as automatic inhibitors for clot formation once a vessel is adequately sealed.

And lastly, an intelligently designed system would account for the inevitable changes in circulation and blood vessel health as organisms age, maintaining efficiency and avoiding clotting-related risks, assuming an intelligent designer was incapable of designing anatomy that doesn't deteriorate with age.

The human blood clotting system, while functional, is far from optimal. Its complexity, vulnerability to errors, and susceptibility to conditions like thrombosis underscore the hallmarks of evolution: a system cobbled together over time, full of trade-offs and inefficiencies. In contrast, a truly intelligent designer could have created a simpler, more reliable system without the risks of excessive clotting or bleeding disorders. The imperfections of the clotting cascade and its frequent malfunctions provide yet another compelling argument against intelligent design and in favour of evolution.

Electrical control of the heart and arrhythmias

The human heart is an extraordinary organ, capable of pumping blood continuously for decades. In the average person it pumps 60-100 times per minute for life. The mammalian heart is

Part II: Internal Weaknesses

composed of two halves separated by a septum, each with two chambers; the atria at the top and the ventricles at the bottom. The right half receives blood from the head and body into the right atrium, which passes it to the right ventricle. This chamber then contracts and forces the blood into the pulmonary circulation where it collects oxygen and gets rid of carbon dioxide. Oxygenated blood returns to the left atrium and into the left ventricle which pumps it out to the head and body in the systemic circulation. This process needs to be precisely coordinated, or it can result in heart failure or an imbalance of pressures in the two circulations, leading to pulmonary oedema and respiratory distress.

Its function is regulated by a sophisticated electrical conduction system, which ensures synchronized contractions of the atria and ventricles. However, this system is prone to a variety of defects, such as Wolfe-Parkinson-White (WPW) syndrome and other arrhythmias.

These malfunctions illustrate the suboptimal, evolutionary nature of the heart's electrical control system rather than the work of an intelligent designer.

First, a brief overview of the heart's electrical system:

The heart's electrical system coordinates the rhythmic contraction and relaxation of the heart muscles, allowing blood to be pumped efficiently. Key components include:

> "Sinuatrial (SA) Node, often called the natural pacemaker. Located in the right atrium, this generates an electrical impulse that initiate a heartbeat. The intrinsic rhythmicity of .cells of the SA node is what sets the heart rate. It can be influenced by hormones such as adrenaline which speeds it up or by acetyl choline, released by the vagus nerve, which slows it down.

Atrioventricular (AV) Node, located in the base of the right atrium, with fibres running down the septum between the two ventricles. The AV node acta as a gatekeeper, delaying the impulse from the atria before it enters the ventricles. This delay ensures the atria contract fully before the ventricles. The fibres that conduct the impulse from the AV node to the ventricles consist of a short 'Bundle of His' which branches in the septum into two sets of Purkinje fibres which take the impulse to the bottom of the ventricles to ensure they contract from the bottom up as a wave of contraction passes up their muscular walls.

This initiation and propagation system ensures that the hearts contractions are coordinated and synchronised.

Cardiac muscle cells are specialised cells with branching junctions between them that ensure a rapid propagation of the electrical impulse.

Contrary to creationists claims and the impression created in the Bible, based on a misunderstanding of why the heart seems to respond to emotions, the heart has no emotional function or ability to know and understand. It is purely a muscular pump whose function is to pump blood.

Bible literalists might like to consider the implications of a book supposedly written or inspired by a creator god, getting the function of the heart and brain confused.

However, for all its seeming perfection, the heart, like other evolved organs and systems has some defects that have their origins in its evolutionary history, and which could have been avoided had the heart been intelligently designed.

Wolfe-Parkinson-White (WPW) Syndrome: This is caused by an accessory conduction pathway (the Bundle of Kent) that bypasses the AV node. This pathway allows electrical impulses to travel between the atria and ventricles too quickly, leading to episodes of rapid heart rate (tachycardia) or other arrhythmias. Symptoms include palpitations, dizziness, fainting, and in severe cases, sudden cardiac death.

Arrhythmias: These occur in two forms: tachycardias (fast) or bradycardia (slow).

By far the commonest is atrial fibrillation[10] (AF): Rapid, chaotic electrical signals cause the atria to contract irregularly, reducing blood-pumping efficiency and increasing the risk of stroke. This can cause an irregular signal being transmitted by the AV node, so the contractions of the ventricles become irregular both is strength and rate. An irregularly irregular pulse at the wrist is a strong indicator of AF.

Ventricular tachycardia is rapidly but coordinated beating of the ventricles, often resulting from damage to the heart from an infarction[11]. The impulse travels more slowly through the damaged muscle so it can emerge after the main contraction has passed, stimulating another wave of contractions so setting up a cycle of rapid ventricular contractions that fail to pump the blood adequately. If left untreated this can quickly become:

Ventricular Fibrillation (VF), which is a breakdown of coordinated contraction/relaxation and effectively cardiac arrest. It can often be reverse with a 'precordia thump' on the breastbone or the delivery of a direct current electric shock via a defibrillator.

---

[10] Fibrillation is where the muscles of the chamber are quivering rather than beating in synchrony. The wave of electrical activity of a normal beat has become uncoordinated and individual fibres as contracting, repolarizing and contracting again, stimulated by an adjacent fibre, not the pacemaker

[11] A blockage in a branch of a coronary artery which deprived the cardiac muscle of oxygen and nutrients.

Bradycardia can be caused by both intrinsic factors such as SA node dysfunction (Sick Sinus Syndrome), myocarditis, cardiomyopathy, delayed conduction through the AV node, and extrinsic factors such as over-stimulation of the vagus nerve by, for example stimulation of the carotid sinus[12], drugs, and medications such as beta-blockers, hypothyroidism[13], hyperkalaemia[14] and hypothermia[15]

Damage to the heart in the area of the AV node or affecting the Bundle of His or the Purkinje fibres, can result in 'Heart block' where the propagation of the electrical impulse is impaired, causing unsynchronized contractions.

Sudden cardiac arrest can occur when an electrical malfunction causes a fatal arrhythmia, especially under stress or in individuals with underlying structural heart issues.

As I mentioned earlier, these problems with the electrical coordination of the heart have their origins in evolutionary compromises.

The mammalian heart evolved incrementally from simpler systems in early vertebrates. Its dual-pump structure (separate atria and ventricles) and conduction pathways are derived from adaptations that worked well in ancestral environments but are

---

[12] A swelling in the carotid artery in the neck in which baroreceptors provide feedback to the cardiac centre to counter a rise in blood pressure by reducing the heart rate. Stimulation of the carotid sinus by a blow to the neck, a bee sting, etc., or massaging the sinus can cause a rapid lowering of the heart rate.
[13] A low level of thyroxine due to an under-active thyroid gland.
[14] An elevated level of potassium in the blood. The correct balance of sodium and potassium in the blood is essential for the activity of 'excitable' tissues such as nerves and muscles.
[15] A low body temperature.

Part II: Internal Weaknesses

prone to failure under modern stresses like prolonged lifespans or sedentary behaviour.

The accessory pathways in WPW syndrome are remnants of embryological development. Normally, these pathways regress during foetal growth, but in WPW, they persist due to incomplete developmental pruning—an error tolerated by evolution because it usually doesn't impact reproductive success.

Primitive hearts, such as those in fish, operate with a single atrium and ventricle, where electrical coordination is simpler. As vertebrates evolved, the transition to a four-chambered heart required more complex electrical pathways to manage the timing of contractions. This added complexity increased the risk of malfunctions.

The heart's system prioritizes rapid conduction (for efficiency) over error-proof mechanisms. This leaves the system vulnerable to abnormal signals, like the re-entry circuits seen in WPW syndrome or atrial fibrillation.

Evolution optimizes traits for reproductive success, not longevity. Many electrical defects, such as atrial fibrillation or conduction blockages, become more common with age, reflecting evolution's indifference to post-reproductive health. The mechanism here is that mutations which improve health in old age are not strongly selected for because they don't affect the number of offspring.

Electrical pathways are delicate and rely on precise signalling. Disruptions, whether caused by genetic mutations, structural abnormalities, or external factors (e.g., stress, caffeine, or medications), easily derail the system.

It is not rocket science to work out how an intelligent designer could have made a better job of designing the heart's electrical control. For example, it could have eliminated redundant pathways such as the Bundle of Kent that causes WPW syndrome instead of keeping it as an atavistic structure in the embryo and sometime failing to suppress its development.

It could have included additional regulatory mechanisms to prevent chaotic electrical activity, such as atrial or ventricular fibrillation, such as automatic "circuit breakers" to stop runaway impulses and restore normal rhythm without external intervention. An internal mechanism for pressing the reset button – which is what defibrillation and cardioversion do.

A simplified conductive pathway could have reduced the risk of blocks, delays, and misfires.

A robust and resilient system would maintain electrical integrity regardless of age, avoiding the increased risk of arrhythmias or conduction blockages seen in older adults, even if the intelligent designer could not avoid designing a body that deteriorates with age

And lastly, our old friend, the failure to anticipate the future and future-proof the design, so a modern lifestyle doesn't conflict with a system optimised for an earlier stage in our evolution.

Yet again, we have a system that is riddled with vulnerabilities that stem from its evolutionary origins. WPW syndrome and other arrhythmias highlight the system's susceptibility to error and its lack of fail-safes, which would be expected in an intelligent design.

Instead, these flaws reflect the trial-and-error nature of evolution, where systems are built incrementally and optimized only for survival and reproduction, not for perfection or long-term functionality. The heart's imperfections are yet another compelling argument against intelligent design and in favour of evolution as the driving force behind the human body's complex but flawed machinery.

Part II: Internal Weaknesses

The pulmonary system

The pulmonary system is sometimes regarded as a separate system of the body, but biologically it is part of the circulatory system as the site of the exchange of gasses with the environment.

As with the other jerry-built systems we've looked at so far, the pulmonary system suffers from the utilitarian, suboptimal solutions and constraints of the evolutionary process and so serves to illustrate the fact there has been no intelligence in the process.

These flaws result in vulnerabilities, inefficiencies, and disease susceptibilities that are difficult to reconcile with the idea of optimal or intelligent design. Below are examples of issues in the pulmonary system that underscore its evolutionary history and limitations:

The first problem is one I mentioned earlier as a shared problem with the digestive system – the vulnerability to choking because the food passages crosses the airway, exacerbated in humans by the lengthening of the pharynx as part of the evolution of complex speech.

This design evolved in vertebrates with less complex systems and was retained as more advanced structures developed. In some species, such as whales, the separation of the trachea and oesophagus is more complete, reducing the risk of choking—a feature humans lack. Even in snakes such as the python which swallows prey several times larger than its head, the top of the trachea can be protruded underneath the object being swallowed so the snake can still breath.

An intelligent design would have been completely separate passages so there is no risk of choking.

But the risk of choking is by no means the only problem with an evolved pulmonary system. It is also vulnerability to infections because it is constantly exposed to airborne pathogens, pollutants, and irritants due to its direct connection to the external environment. This leads to conditions like pneumonia, bronchitis, and tuberculosis. Of all the body's systems the pulmonary system is probably unique in providing almost ideal conditions for opportunistic bacteria to set up home. It is warm, moist, and highly vascular.

It evolved this way because of the need for constant air intake to supply the circulatory system with oxygen to supply cells that are remote from the environment, unlike the cells of single-celled organisms. But there's no mechanism to entirely avoid inhalation of harmful particles, such as you might expect an intelligent designer to have provides.

The efficient functioning of the lungs relies on thin-walled air sacs, or alveoli surrounded by capillaries. While efficient under normal conditions, this design is highly vulnerable to pulmonary oedema, where fluid accumulates in the alveoli and interferes with oxygen exchange.

Because the efficient exchange of gasses depends on the partial pressure[16] of the sir in the alveoli, humans are restricted in the altitudes they can live in.

Prolonged living at high altitudes stimulates the body to produce more red blood cells to compensate for the lack of oxygen. This

---

[16] Partial pressure is a function of the absolute pressure divided by the ratio of gasses in the atmosphere. At high altitudes the problem is not, as is commonly believed, a lack of oxygen, which remains at the same percentage in the atmosphere, but a reduction in air pressure which results in a low partial pressure of oxygen (PO_2). Raising the proportion of oxygen in respired air can restore the PO_2 to a physiologically acceptable level.

Part II: Internal Weaknesses

causes the blood to thicken leading to longer term problems known as altitude sickness. It has recently been discovered that Tibetans can live at the high altitudes of the Tibetan Plateau because they have a gene inherited from Denisovans (21).

The lungs evolved for environments close to sea level. Modern humans' ability to travel rapidly to high altitudes or survive long-term requires adaptations that the evolutionary process has not yet caught up with. An intelligent design process could have anticipated and planned for this requirement or produced a more flexible and adaptable respiratory system.

Respiration in humans relies on the proper functioning of the diaphragm. However, the diaphragm is a muscle prone to injury, herniation, and inefficiency under certain conditions and evolved in earlier vertebrates as shown by hiccups. Hiccups are reflexive spasm of the diaphragm which serves no known purpose but are probably a vestige from amphibian ancestors that relied on gulping air.

The diaphragm is also vulnerable to paralysis caused by damage to the phrenic verve[17]. The diaphragm evolved from simpler respiratory systems in early vertebrates. While effective, its dependence on a single nerve (phrenic nerve) makes it vulnerable.

---

[17] The phrenic nerve is a mixed motor/sensory nerve that originates from the C3–C5 spinal nerves in the neck. The nerve is important for breathing because it provides exclusive motor control of the diaphragm, the primary muscle of respiration. In humans, the right and left phrenic nerves are primarily supplied by the C4 spinal nerve, but there is also a contribution from the C3 and C5 spinal nerves. From its origin in the neck, the nerve travels downward into the chest to pass between the heart and lungs towards the diaphragm.(Wikipedia)

A more distributed or redundant muscle and nerve system for breathing would prevent complete dependence on a single nerve or muscle.

The bronchi[18] and bronchioles[19] are prone to narrowing and obstruction due to inflammation, excess mucus, or structural collapse, leading to diseases such as, asthma:

Asthma is itself the result a design flaw with an over-sensitive immune system and is a hypersensitive reaction of the airways that can lead to life-threatening constriction.

Chronic Obstructive Pulmonary Disease (COPD) includes chronic bronchitis and emphysema, which impair airflow and gas exchange. The problem arises because the branching structure of the airways maximizes surface area for gas exchange but increases resistance to airflow. Evolution prioritized surface area over robustness against obstruction.

An intelligent designer could have designed larger or more rigid airways that resist collapse and constriction, without compromising gas exchange efficiency.

At times of increased demand or stress the human pulmonary system struggles to meet oxygen demands, leading to hyperventilation. This overcompensation for oxygen needs, can cause dizziness or fainting.

The syndrome known as hyperventilation is a result of the way the level of carbon dioxide in the blood is measured by chemoreceptors that communicate with the respiratory centre in the Medulla. As CO_2 levels rise, so respiration increases in rate and depth and as it falls, so the rate and depth of respiration fall.

---

[18] The main branches that lead from the bottom if the trachea to the left and right lungs.
[19] The finer branches of the airway that lead from the bronchi and terminate in the alveoli or air sacs where gas exchange takes place.

Part II: Internal Weaknesses

However, the rate of respiration can also be influenced by higher centres in the brain so can be increased voluntarily or due to anxiety, as a primitive preparation for fight or flight.

Continued hyperventilation flushes more CO_2 out of the blood which depresses the breathing reflex and changes the acidity of the blood where CO_2 is part of the pH regulation mechanism. This increase in acidity can cause the muscles of the face, arms, and hands to go into a spasm and the patient's anxiety increases even more. They commonly believe they are dying. The simple remedy is to persuade them to breathe into and out of a paper bag held over the mouth. This rebreathing raises the CO_2 in the inspired air which raises it in the blood, reversing the syndrome, usually in a few minutes.

In my former life as a paramedic I lost count of the number of times I performed this 'miracle cure' on a patient, even on one occasion, a doctor.

As an adolescent at school, I was something of an athlete, especially a sprinter and held a number of school record for running and jumping. Running the final leg of a 4 x 110-yard relay race, I had taken the baton in last place and passed the first two opponents within the first 50 yards, leaving only the leading sprinter to overtake. Coming off the final bend I was about 10 yards behind with about 30 yards to the tape. So determined was I to win that I forgot to breathe and passed him with about 5 yards to spare, and my legs turned to jelly as I hit the tape. I lay on the ground struggling to breathe.

My body had gone into oxygen debt having burned up the reserve in my muscles and resorted to anoxic respiration that produced lactic acid in my muscles that were now cramped and painful. My muscles were demanding payment for the oxygen I had borrowed from them, and I couldn't shift enough air in an out of my lungs to supply the demand.

For several minutes I thought I was dying as the faces peering down at me swirled in a blur, until the debt was repaid and normal service to my brain was resumed. I learned then that the key to survival is to control your breathing. Had I been running from a leopard on the African savannah, I would have been an easy meal and, more importantly in evolutionary terms, left no descendants, unless I had outrun my companions, in which case my fast twitch fibres would have been inherited by my children, unless the leopard also had a companion.

The system has evolved for moderate levels of activity, typical of hunter-gatherer lifestyles. Modern athletic or stressful conditions push the system beyond its natural capacity. An intelligent designer could have provided a more adaptable oxygen delivery system, such as auxiliary oxygen storage or more efficient oxygen-carrying molecules could improve performance and reduce strain.

Compared to the respiratory system of birds, the mammalian respiratory system is primitive to say the least. Birds are capable of prolonged periods of sustained flight, even at altitudes at which a human would lose consciousness without a supply of oxygen.

At birth, the lungs inflate successfully at the first breath because surfactant molecules prevent the alveolar walls from adhering. But the lungs of premature infants are underdeveloped, lacking these surfactants to keep alveoli[20] from collapsing. This can lead to neonatal respiratory distress syndrome (RDS), which is life-threatening without medical intervention. The problem is, human gestation evolved to balance pelvic constraints with brain development, leading to births before full organ maturity in

---

[20] The air sacs in which exchange of gasses takes place.

many cases. The solution to this would not be beyond the wit of even a half-wit designer, let alone an intelligent designer.

One of the consequences of an upright posture is that it increases the risk of atelectasis (lung collapse) in the lower lobes due to gravity compressing the lungs when lying down. Similarly, fluids can pool in lower lung areas, increasing the risk of infections or complications in bedridden patients. The problem is that the transition from quadrupedal to bipedal posture did not allow for a complete reorganization of the pulmonary system to optimize for vertical orientation. This is one way we can tell that bipedalism wasn't intelligently designed, which would have entailed planning for just these sorts of complications.

Their exposure to carcinogens in the air such as tobacco smoke and industrial pollutants makes the lungs especially vulnerable to cancers. When we were preindustrial mammals this was not something that an evolutionary process could anticipate or plan for, bur an intelligent designer could have produced a more robust respiratory system complete with a repair mechanism that could have coped with the genetic damage pollutants can cause, so reducing the risk of lung cancers. I'll have more to say on the subject of cancers and how they could have been avoided in a later section.

The pulmonary system, like the rest of the human body, demonstrates a mix of functionality and vulnerability. Its flaws—shared pathways with the digestive system, susceptibility to infections, and inefficiencies in gas exchange—point to the incremental and imperfect process of evolution. These imperfections align with what we would expect from a system shaped by natural selection rather than by an omnipotent, intelligent designer. The pulmonary system works well enough to sustain life but is far from the optimal design that might be envisioned by a truly intelligent creator.

Varicose veins

Varicose veins—swollen, twisted veins that are often visible under the skin, particularly in the legs—are a striking example of how evolutionary constraints and adaptations can result in a suboptimal system. These veins reflect the imperfections inherent in human physiology and provide a compelling argument against intelligent design. Below, we examine the causes, evolution, and implications of varicose veins in the context of the debate between evolution and intelligent design.

First a definition. Varicose veins occur when veins, especially in the legs, become enlarged and their valves fail to function properly. This leads to blood pooling and the characteristic swollen, twisted appearance. This causes pain, swelling, heaviness in the legs, skin discoloration, and, in severe cases, ulcers or blood clots (thrombophlebitis).

The underlying cause is that veins rely on one-way valves to keep blood flowing toward the heart. When these valves weaken or fail, blood flows backward (venous insufficiency), causing the veins to stretch and become dysfunctional. The pressure from the arteries and the suction of the venous return to the heart have to fight against gravity, aided only by the muscular contractions of the leg muscles such as the calf muscles during exercise and the one-way valves in the veins.

Varicose veins illustrate a mismatch between evolutionary pressures and modern human physiology, as well as inherent flaws in our circulatory system. Humans are the only mammals that walk fully upright. The transition from quadrupedal to bipedal posture increased the gravitational burden on the veins in the lower body, particularly in the legs. Blood must now work against gravity to return to the heart. Consequently, this evolutionary adaptation led to increased pressure in the leg veins, which makes valve failure more likely over time.

Part II: Internal Weaknesses

Any intelligent designer worthy of the name would foresee this issue and engineer a more robust venous system to accommodate upright posture, such as, veins with thicker walls to resist stretching, and/or a fail-safe mechanism for valve function.

The one-way valves in veins are relatively simple and fragile structures, being a near-enough-is-good-enough solution to a problem in an ancestral mammal. However, over time, they are prone to mechanical failure, particularly under the increased strain of standing or prolonged sitting, consequently, valve failure leads to retrograde blood flow, increased venous pressure, and the formation of varicose veins.

The venous system relies on a single valve per segment to prevent blood backflow. If one valve fails, the entire segment of the vein can become dysfunctional.

A more critical system might provide sufficient evolutionary pressure to evolve a more robust, fail-safe system, however, as with so many other design flaws in human anatomy and physiology, evolution has prioritized reproduction over longevity and comfort in old age, so problems that develop in later life are given a low priority.

The hunter-gatherer hominin in which bipedalism evolved would probably have been almost constantly on the move when not sleeping, so evolutionary adaptations tended to be optimised for an active existence and possibly several miles of walking a day.

That evolutionary optimization is clearly not optimal for today's comparatively sedentary lifestyle. Now prolonged standing or sitting, obesity, and lack of exercise worsen venous pressure, contributing to the development of varicose veins.

The complications of varicose veins further underscore their poor design: chronic venous insufficiency can lead to debilitating discomfort and ulceration, when pooling blood can cause skin breakdown and venous ulcers, which are difficult to heal. Additionally, blood clots can form in varicose veins, potentially leading to life-threatening deep vein thrombosis (DVT), which can result in loss of a limb and a fatal pulmonary embolism if a clot breaks free and passes through the heart into the pulmonary circulation.

Varicose veins are the result of the incremental adaptations to bipedalism that we have seen before with other systems. The trade-off was between the advantages of walking upright which freed the hands and allowed for greater mobility and tool use, which offered significant evolutionary advantages, and the mechanical demands on the venous system, leading to increased failure rates in lower-body veins.

As I've said before, evolution optimizes for survival and reproduction, not perfection. Veins and valves that work well enough to maintain circulation through most of an individual's life were favoured, even if they fail later. Evolution is constrained by having to make do with existing structures and can't start from scratch as an intelligent designer could. In this case the human venous system evolved from a design better suited to quadrupeds and retained many of its limitations.

Intelligent design advocates argue that the human body was purposefully created by a superior intelligence. However, the prevalence and dysfunction of varicose veins are hard to reconcile with this view, since an intelligent designer could easily have implemented stronger valves, more efficient venous systems, or alternative methods for blood return to the heart.

The fact that varicose veins are so common—affecting up to 30% of adults globally—suggests a lack of foresight or care in the "design.", so we can safely conclude then that varicose veins are a clear example of a flawed system shaped by evolutionary pressures rather than intelligent design. Their prevalence, mechanical vulnerability, and complications highlight how evolution produces systems that are functional but far from perfect. These imperfections make sense when viewed through the lens of natural selection but are difficult to justify under the premise of a perfect, intelligent creator. Instead, varicose veins underscore the evolutionary principle of trade-offs and the constraints of modifying preexisting structures.

Coronary circulation and heart attacks

The coronary circulation—comprising the network of blood vessels that supply oxygen and nutrients to the heart muscle—provides critical support for cardiac function. However, the system has significant vulnerabilities and inefficiencies that point to its evolutionary origins rather than the work of an intelligent designer. Below, we explore how the coronary circulation reflects evolutionary trade-offs and constraints, often to the detriment of human health.

The coronary arteries deliver oxygenated blood to the myocardium (heart muscle). The main vessels include the left coronary artery (LCA), which divides into the left anterior descending artery (LAD) and the circumflex artery, and the right coronary artery (RCA). Blockages (e.g., due to atherosclerosis) in these arteries can lead to myocardial infarction (heart attack), often with fatal consequences.

The significant flaws in the coronary artery design are:

They are 'end arteries' with no redundancy, meaning they have minimal collateral circulation. If a coronary artery is blocked, the region of the heart it supplies suffers ischemia (oxygen deprivation) and dies if blood flow is not restored.

This potentially life-threatening design flaw probably evolved in a simpler organism with lower oxygen demands. Over time, as hearts became more complex, the design persisted without substantial improvement in redundancy. An intelligent designer could have provided more redundancy, so a blockage could be bypassed.

Coronary arteries are also susceptible to atherosclerosis - the buildup of fatty plaques, which can narrow or completely block blood flow. Together with the coronary arteries being 'end arteries' this is a major design flaw which is a frequent cause of death.

Risk factors include diet, genetics, smoking and inflammation.

The evolutionary cause of this situation is that humans evolved in environments with limited food availability. This favoured genes that store energy efficiently. In modern environments with abundant high-fat diets, these once-beneficial traits contribute to atherosclerosis. An intelligent designer could have devised a system which didn't cause plaque to build up on the walls of arteries supplying critical systems. One might be forgiven for thinking this system was designed to cause sudden death, but of course, we know it evolved by a system which optimises in favour of reproduction, not longevity, so sudden death in middle to later years carries a low evolutionary priority.

Compared to arteries supplying blood to other organs, coronary arteries are relatively narrow. This increases the risk of a fatal blockage. This evolved to meet the demands of early hominins with less physical strain. Modern humans' increased stress levels, sedentary lifestyles, and longer lifespans exacerbate the limitations of this design. Larger blood vessels or an auxiliary supply would have solved these problems, even if the build-up of plaque couldn't be avoided.

The heart is also vulnerable to circulatory failure because it receives blood during diastole, when the heart is relaxed, creating a dependency on adequate diastolic pressure. If diastolic

Part II: Internal Weaknesses

pressure drops (e.g., in shock or severe blood loss), coronary perfusion is compromised, risking heart failure. Although this diastolic flow dependency maximizes oxygen delivery during the heart's relaxation phase, it leaves the system vulnerable to blood pressure fluctuations. A more robust system would have ensured a continuous blood flow during systole and diastole.

The heart is disproportionately dependent os a single artery – the Left Anterior Descending Artery, also called the "Widowmaker". The LAD artery supplies a significant portion of the left ventricle. Blockages in this artery are often fatal, earning it the nickname "the widowmaker."

The LAD evolved to supply the largest and most critical portion of the heart. However, this evolutionary optimization came at the cost of high stakes—failure in this artery has devastating consequences. Not quite having all your eggs in one basket, but close. You have all your life in one artery. I defy anyone to explain how this is the work of an intelligent designer.

Creationists would have us believe that an intelligent designer would design something that:

> Is a leading cause of death worldwide. Heart attacks occur when blood flow in a coronary artery is blocked. The heart's lack of regenerative capacity exacerbates the damage?

> Reflects poor design in that plaque buildup in coronary arteries progresses silently over decades, often only becoming symptomatic during exertion or stress.

> Is so inefficient that reduced blood flow in coronary arteries causes chest pain or angina during physical or emotional stress?

The evolutionary explanation is that coronary circulation evolved incrementally, constrained by the designs of ancestral vertebrates. In fish, oxygenated blood from the gills flows directly to the heart. As vertebrates transitioned to land, coronary arteries evolved to meet the higher oxygen demands of terrestrial organisms.

Mammals developed a four-chambered heart with a coronary system that could sustain their higher metabolic rates.

Evolution prioritized a system that works "well enough" for survival and reproduction, typically until the end of reproductive age. Coronary diseases often manifest later in life, after reproductive success is achieved, making them less relevant to evolutionary pressures.

Evolution works with existing structures, modifying them over time. The coronary system's vulnerabilities reflect these historical constraints rather than optimal planning.

Proponents of intelligent design claim the human body is the result of purposeful creation. However, the coronary circulation poses challenges to this view:

> Why would an intelligent designer create a system so vulnerable to blockages and failure?
>
> Why is there no built-in redundancy to compensate for coronary artery dysfunction?
>
> Why design arteries that are prone to life-threatening diseases like atherosclerosis, particularly given the prevalence of modern risk factors?

These questions underscore the incompatibility of the coronary system's flaws with the notion of a perfect creator. The coronary circulation exemplifies the evolutionary process: a system that is functional but far from perfect. Its flaws, such as susceptibility to blockages, lack of redundancy, and dependency on diastolic blood flow, make sense in the context of natural selection but are

Part II: Internal Weaknesses

inexplicable under the framework of intelligent design. These imperfections highlight the "good enough" nature of evolution, contrasting starkly with the optimal solutions an intelligent designer might have implemented. The coronary system's vulnerabilities are a powerful argument in favour of evolution and against intelligent design.

Part II: Internal Weaknesses

The reproductive systems

The human reproductive system is a complex but deeply flawed result of evolutionary tinkering. While it serves its fundamental purpose of perpetuating the species, its numerous inefficiencies, vulnerabilities, and unnecessary complexities highlight its evolutionary origins and the lack of any intelligent involvement.

First, I'll deal with the male reproductive system which, though not so complex and error-prone as the female reproductive system nevertheless has some glaring faults that defy any notion of intelligent design.

The Male Reproductive System

The vas deferens.

The vas deferens, the duct that carries sperm from the testes to the urethra, takes a long and convoluted path, looping around the ureters (which carry urine from the kidneys to the bladder) instead of following a direct route.   This can give rise to the condition known as a testicular torsion[21].

During early vertebrate evolution, the testes formed inside the body. As mammals evolved, the testes migrated outside the body to maintain a lower temperature for optimal sperm production, but the vas deferens retained its original pathway.

Any intelligent designer worthy of the name could have arranged a direct, efficient path for sperm transport to eliminate this

---

[21] Testicular torsion occurs when the spermatic cord (from which the testicle is suspended) twists, cutting off the blood supply to the testicle. The most common symptom in children is sudden, severe testicular pain. If left uncorrected for 12 hours or more, it can result in the loss of the testicle. About 40% of cases result in surgical removal of the testicle.

redundancy and reduce the risk of complications like testicular torsion.

Testicular Vulnerability.

In males, the testes are housed outside the body in the scrotum, making them highly vulnerable to trauma and temperature fluctuations.

The reason for this is that sperm production evolved and was optimized for cold-blooded vertebrates. Positioning them in a skin flap outside the abdominal cavity is a Heath-Robinson suboptimal workaround for the problem caused by the beneficial evolution of warm-bloodedness.

Intelligently designed sperm production would not need this shoddy fix because it would work for testes housed inside the body the same as the ovaries are. In fact, it works perfectly well in elephants and other Afrotheria such as manatees, so there appear to be no insurmountable problems in doing so.

Prostate Gland Issues

The prostate gland surrounds the urethra, and as men age, it often enlarges (benign prostatic hyperplasia), causing urinary obstruction, and increasing the risk of urinary tract infections (UTIs). Urinary retention is a painful condition requiring the insertion of a catheter to provide passage for urine, with the subsequent loss of muscular control of urination and a risk of urinary tract infection (UTI).

Prostate cancer is also one of the most common cancers in men.

The evolutionary explanation is that the prostate evolved as an accessory gland to enhance sperm survival in the female reproductive tract. However, in an example of how evolution optimizes for reproduction, not long-term post-reproduction survival, there was little evolutionary pressure to reposition the gland or prevent it enlarging with age. How an intelligent designer could have done better is probably too obvious to spell out, unless urinary retention and prostate cancers were part of its divine plan.

The Female Reproductive System

Birth canal and pelvic evolution

The problems evolutionary changes to the pelvis have caused in respect of the birth canal in the course of evolving bipedalism and an upright posture were described in Part I.

Other major flaws, how they evolved and why any intelligent designer could have made a better job of them, are:

Ectopic Pregnancies

In some cases, a fertilized egg implants outside the uterus, typically in the fallopian tubes. These pregnancies are almost always nonviable and life-threatening, and usually result in the loss of the fallopian tube on the affected side, reducing fertility.

The evolutionary reason can be found in our ancestry when fallopian tubes were part of an earlier reproductive system in vertebrates. The lack of a precise mechanism for ensuring implantation in the uterus reflects the constraints of evolution. An intelligent solution would be to ensure the zygote moves quickly to the uterus before implanting or is only fertilised in the uterus. This would abolish the risk of an ectopic pregnancy as would making it impossible for the zygote to embed anywhere but in the uterus.

Placenta previa

As well as embedding outside the uterus, the zygote can also embed low down in the uterus, so the placenta is close to or even covering the cervix. Delivering the baby through the cervix with a placenta previa or even allowing the pregnancy to go to full term and risk the beginnings of labour, carries a high risk of a catastrophic haemorrhage endangering the lives of both mother and baby, so the usual procedure is a Caesarean delivery (C-section) during the 8th month of pregnancy.

Placenta previa in third-world economies, with poor antenatal services is a significant cause of perinatal mortality.

The problem of placenta previa arises because:

> In humans, the placenta is haemochorial, meaning it deeply invades the endometrium to establish a robust connection with the maternal blood supply. This aggressive implantation strategy evolved to meet the high nutritional demands of the large human brain. However, the invasive nature of the placenta increases the likelihood of improper implantation, such as in the lower uterus (placenta previa). This reflects a trade-off between evolutionary adaptations for foetal nourishment and the structural constraints of the uterus.
>
> The evolution of bipedalism narrowed the human pelvis, changing the shape and size of the uterus. This may have increased the chances of abnormal placental positioning, as the placenta has limited "real estate" for proper implantation.
>
> The uterus evolved to support pregnancies in quadrupeds, where gravity and uterine shape reduced the risks of placenta previa. These constraints carried over into bipedal humans without significant redesign.

If an intelligent designer created the human reproductive system, one would expect:

> A mechanism to ensure that the placenta implants only in the upper uterus, by, for example, restricting the endometrium to 'safe' parts of the uterus.

Part II: Internal Weaknesses

A less invasive placental attachment to reduce the risk of complications. The occurrence of placenta previa is entirely inconsistent with the notion of intelligent design by an omnibenevolent creator.

Pre-eclampsia.

Pre-eclampsia is a pregnancy complication characterized by high blood pressure and damage to organ systems, often the kidneys or liver. It typically occurs after 20 weeks of pregnancy and, in severe cases, can be life-threatening for both mother and baby. Looked at dispassionately, a developing foetus resembles a parasitic organism developing inside a host.

Pre-eclampsia is thought to result from an evolutionary "arms race" between the mother and foetus, just as we frequently see between parasite and host. The foetus demands resources (e.g., nutrients and oxygen) from the mother, while the mother's body attempts to regulate resource allocation to preserve her own health.

Pre-eclampsia arises when the placenta releases signals that increase maternal blood pressure to improve nutrient delivery, but this adaptation can overshoot and harm the mother.

The human placenta's deep invasion of maternal tissues increases the risk of maternal health complications, including pre-eclampsia. This aggressive strategy likely evolved to support the brain growth of human offspring but at the cost of increased risks to maternal health.

This arms race between a supposedly intelligently designed mother and a baby supposedly designed by the same intelligent designer is of course, utterly inconsistent with the notion of an intelligent designer since it is the epitome of stupidity to indulge in an arms race essentially with oneself, to no ultimate benefit to either the mother or baby and possible harm to both.

Evolution shaped the human reproductive system in an environment where caloric resources were scarce. Modern diets and lifestyles, combined with longer life expectancies, exacerbate the risks of pre-eclampsia and other pregnancy-related complications.

An intelligently designed reproductive system would be expected to:

> Avoid the maternal-foetal conflict by balancing the needs of the foetus and mother.

> Prevent pre-eclampsia by ensuring proper vascular remodelling and blood pressure regulation. The existence of pre-eclampsia, with its severe risks, suggests a system that evolved through trial and error rather than deliberate intelligent design.

The evolutionary struggle between maternal and foetal interests is a key driver of pregnancy complications, such as pre-eclampsia and placenta previa. This conflict undermines the notion of intelligent design, as it produces inefficiencies and risks that are incompatible with the idea of a benevolent and omniscient designer.

The human reproductive system lacks safeguards against conditions like placenta previa and pre-eclampsia. A well-designed system would include mechanisms to prevent improper placental implantation or excessive placental invasion.

These risks and inefficiencies of the human reproductive system strongly support the argument for evolution and undermine claims of intelligent design.

Part II: Internal Weaknesses

Menstruation

Human females shed the uterine lining (endometrium) during menstruation if no pregnancy occurs, resulting in blood loss and vulnerability to infection. It also places an additional demand for iron to replace the lost blood, meaning women need about twice the dietary iron intake of men, making post-pubertal women susceptible to iron-deficiency anaemia.

Menstruation is believed to have evolved as a mechanism to prevent pathogens from colonizing the uterus. However, this process is energetically costly and not shared by most mammals, reflecting its evolutionary contingency.

There may have been additional evolutionary pressure to evolve this 'cleansing' when human sexual activity became dissociated from reproduction and took on a recreational, social bonding role. Humans are unusual among mammals in their pattern of sexual behaviour:

> Unlike many animals that engage in sexual activity only during oestrus (when females are ovulating), human females do not exhibit clear signs of fertility and are sexually receptive throughout their cycle.

> The decoupling of sex from reproduction allows for more frequent intercourse, which facilitates pair bonding and social cohesion but also increases the risk of introducing pathogens into the female reproductive tract.

Menstruation may have evolved as an adaptive mechanism to mitigate these risks.

> By periodically clearing the uterine lining, menstruation may act as a form of immune defence, flushing out pathogens introduced during sexual activity.

The link between frequent sexual activity and pathogen exposure could explain why humans, unlike most mammals, menstruate visibly and regularly.

Human sexuality also plays a significant role in social bonding and group cohesion, which may have influenced the evolution of menstruation.

In ancestral human societies, sexual activity likely occurred within groups with varying degrees of monogamy and polygamy. Increased sexual activity among group members would elevate the risk of uterine infections. In such a cultural environment, menstruation could have provided an additional layer of defence, allowing females to maintain reproductive health while engaging in frequent sexual activity. This aligns with the idea that menstruation is not just a side effect of reproductive biology but a potential evolutionary adaptation shaped by human mating systems and social behaviours.

It may upset religious fundamentalists, who still insist the purpose of sexual intercourse is procreation and should be restricted to married couples to learn that, if menstruation had been intelligently designed, it might have been to facilitate recreational and polygamous sex.

However, there are compelling reasons to think that it evolved with no intelligent involvement anywhere in the process because it is entirely inconsistent with the whole notion of intelligent design.

Menstruation is energetically costly and increases vulnerability to anaemia and infection. However, these costs may have been outweighed by the benefits of maintaining uterine health in a species with high rates of sexual activity. The trade-off suggests that menstruation evolved in the context of balancing reproductive fitness with the risks posed by frequent intercourse.

Part II: Internal Weaknesses

The evolution of menstruation as a means of pathogen defence fits well within the broader framework of human sexual and social behaviour. Humans' unique pattern of extended sexual receptivity and frequent intercourse, including with multiple partners, likely increased the need for mechanisms to maintain female reproductive tract health. Menstruation, while costly, may have evolved as an adaptive response to these pressures, highlighting the complex interplay between reproduction, sexuality, and social dynamics in human evolution. This connection also underscores the imperfect and patchwork nature of evolution, as menstruation comes with its own set of drawbacks and inefficiencies.

Since most mammals manage perfectly well without a cyclical breakdown of the endometrium, there appears to be no good reason for human females to menstruate, but if there is a compelling reason why the endometrium needs to be replaced every 28 days, give or take a day or two, a system that reabsorbs it or avoids its cyclical buildup would conserve resources and reduce risks and it would not have been beyond the wit and capabilities of an intelligent designer to produce such a solution.

Hyperemesis gravidarum (severe morning sickness).

Hyperemesis gravidarum (HG) is a severe form of nausea and vomiting during pregnancy, which highlights the contrast between what evolution has produced and what an intelligent design process could have produced

While its exact causes are not fully understood, HG may reflect evolutionary trade-offs associated with pregnancy, offering potential adaptive benefits while also exposing vulnerabilities that refute any notion of intelligent design.

HG is characterized by:

    Severe and persistent nausea and vomiting.

Significant weight loss (>5% of pre-pregnancy weight).

Dehydration and electrolyte imbalances.

In extreme cases, it can require hospitalization and may endanger both maternal and foetal health.

It is distinct from the more common "morning sickness" experienced by most pregnant women and is far more debilitating.

HG may be an exaggerated form of a beneficial evolutionary adaptation. Pregnancy-related nausea and vomiting are thought to have evolved to protect the developing foetus during its most vulnerable stage (the first trimester), when organ systems are forming and the risk of teratogenic (foetal-damaging) harm is highest. It helps the mother avoid foods that could harm the foetus, such as those containing toxins, pathogens, or harmful secondary compounds. For example, foods with high bacterial loads such as raw meat and eggs, or bitter or spoiled foods, which may indicate toxins.

Hormones like human chorionic gonadotrophin (hCG), which rise dramatically during early pregnancy, may play a dual role in promoting nausea and supporting the development of the placenta. The high hCG levels observed in HG could represent an exaggerated response to an otherwise adaptive process.

However, in HG, this adaptation has become maladaptive and the severity of the symptoms far outweighs any protective benefit, resulting in potentially life-threatening conditions for both mother and foetus.

The severity of HG can also be seen as an arms race between competing evolutionary pressures:

Human pregnancy involves intense competition between the foetus and mother for resources. The foetus "demands" nutrients via hormonal signals, while the mother's body attempts to regulate resource allocation to protect her own health. In HG, this balance may be tipped excessively in favour of the foetus, with elevated hCG levels and other hormones leading to extreme symptoms in the mother.

The evolution of bipedalism altered the shape and position of the human uterus, potentially making some aspects of pregnancy more taxing. The changes in pressure within the abdominal cavity and the displacement of other internal organs is different in humans than it was in ancestral mammals. While this may not directly cause HG, it adds to the complexity of human pregnancy.

As with so many other problems with human anatomy and physiology there is a mismatch between what evolution optimised for on the African savannah and what is needed today.

In ancestral environments, women experiencing nausea and vomiting may have been more likely to avoid potentially harmful foods in environments where food contamination was common.

In hunter-gatherer societies, there might have been a better social safety net so women suffering from severe symptoms might have been supported by the group, mitigating the risks of HG. In modern societies this group support may be missing.

It is inconceivable that a benevolent, intelligent designer would have designed HG, as it seems unnecessarily harmful and inefficient. It can lead to extreme malnutrition or dehydration; it can jeopardize both maternal and foetal survival. A well-designed system would regulate nausea and vomiting to avoid harm while still achieving protective benefits.

HG represents a system that goes haywire without any clear failsafe and the extreme rise in hCG and its connection to HG underscores the "messy" nature of human biology, where evolutionary processes often optimize for short-term reproductive success at the expense of individual well-being.

It's safe to say then that hyperemesis gravidarum, like many other complications of human pregnancy, reflects the evolutionary constraints and trade-offs inherent in our biology. While nausea and vomiting during pregnancy likely evolved as an adaptive response to protect the foetus, HG represents an exaggerated and maladaptive version of this process. The condition underscores the imperfect, often messy outcomes of evolution, challenging claims of intelligent design. A truly "intelligent" reproductive system would balance foetal protection with maternal well-being, avoiding the extreme suffering and risks posed by HG, and a pregnancy would more closely resemble a symbiotic association rather than the parasitic arms race between the foetus and the host mother.

The casual dismissal of the problems associated with childbirth as the result of 'sin' inherited from Eve, as Professor Behe does with his 'devolution' nonsense is a form of victim blaming unworthy of people who profess to be kind, caring and compassionate. As though a pregnant woman today can have any influence or share any guilt for the alleged actions of a character in an origin myth that was arbitrarily declared to be an inherited sin in the fearful infancy of our species in the Late Bronze Age. It is as absurd and heartless as declaring that all women bear a share of the guilt for something done by Helen of Troy or Aphrodite.

It also destroys any claim that creationism is real science, not Christian fundamentalism in disguise.

Menopause

Unlike most animals, human females experience menopause, a prolonged post-reproductive phase associated with hormonal

Part II: Internal Weaknesses

imbalances and increased risk of osteoporosis and cardiovascular disease. It might have evolved, according to the "grandmother hypothesis", to allow older women to assist with raising grandchildren, enhancing the survival of their genes. However, this explanation reflects a compromise rather than an optimized design.

Incidentally, the fact the women remain sexually active after the menopause, with some even experiencing increased libido, runs counter to the religious fundamentalist claim that the purpose of sexual intercourse is procreation and anything else is some sort of 'sin'. In post-menopausal women it clearly has a different role, more akin to pair-bonding with a partner or social cohesion.

The evolutionary explanation of additional support for children with children of their own, while free from pregnancy and childbirth, is compelling, whereas the notion that an intelligent design would result in sexually active women being infertile while declining in the capacity even to be 'grandmothers' makes as much sense as a back pocket in a vest.

Reproductive Cancers

Cancers of the reproductive organs (e.g., testicular, ovarian, cervical, and prostate cancers) are common and often fatal. They arise because rapid cell division in the reproductive tissues increases the likelihood of mutations. Evolution has prioritised reproductive success, over long-term survival, sacrificing our future welfare for the sake of producing more children.

An intelligent designer could have designed a more robust, less error-prone DNA replication process or provided better error-checking and repair.

The human reproductive system provides compelling evidence for evolution and against intelligent design. For example, it is heavily influenced by its evolutionary history, including structures and pathways inherited from distant vertebrate ancestors.

Evolutionary trade-offs have produced pelvic narrowing for bipedalism and a curved pelvis for an upright posture, at the expense of safe childbirth.

Arms races between the mother and the 'parasitic' foetus have predisposed women to pre-eclampsia and placenta previa.

These compromises lead to suboptimal outcomes. Evolution cannot "start over" but must work with existing structures, leading to inefficient and flawed systems like the convoluted vas deferens or the high-risk birthing process.

Intelligent design advocates argue that the human reproductive system reflects purposeful creation. However:

> The high risks associated with childbirth, ectopic pregnancies, and reproductive cancers challenge the notion of a benevolent, omniscient designer.
>
> The inefficiencies, redundancies, and shared developmental origins of male and female systems suggest a patchwork process rather than deliberate planning.

The human reproductive system is a prime example of evolutionary compromise and constraint, riddled with inefficiencies and vulnerabilities. While it performs its essential function, its flaws and risks are difficult to reconcile with the idea of an intelligent designer. Instead, the reproductive system highlights the trial-and-error nature of evolution, where "good enough" is the rule rather than perfection.

Part III: Inefficient or Wasteful Systems

The Body of Evidence

Part III: Inefficient or Wasteful Systems

The Energy Hogging Brain

Although the human brain is possibly the most amazing examples of what the process of evolution has achieved, it is also one of the least efficient and energy-intensive systems, because the process of evolution has prioritized function over efficiency.

Its relatively large size has necessitated the skeletal changes we met in the earlier section, with all the problems inherent in remodelling the face. There are increased risks in pregnancy with the developing foetus's demands for nutrients to build a big brain conflicting with the mother's own needs, and the attendant difficulties in giving birth to a baby with a large head through a birth canal made narrow and curved by modifications for bipedalism and an upright posture.

The human brain's high energy demands are a striking example of inefficiency that challenges the concept of intelligent design. The brain accounts for only 2% of total body weight but consumes approximately 20% of the body's energy at rest. This energy is primarily used for:

Maintaining neuronal ion gradients[22] (critical for transmitting electrical signals) and needing a constant supply of oxygen and glucose.

Supporting synaptic activity, which underpins learning, memory, and cognition.

Basic metabolic processes to keep brain cells alive.

---

[22] When a neuron 'fires' sodium ions flood into the cell causing the cell membrane to depolarize. These must then be actively pumped out of the cell using the neuron's energy supply which needs to be restored by the mitochondria using oxygen and glucose to convert ADP to ATP.

The Body of Evidence

The brain relies heavily on glucose as its primary energy source, but unlike other organs, it cannot store significant amounts of glucose and depends on a constant supply from the bloodstream.

During periods of fasting, the brain adapts to use ketones, but this is less efficient and can impact cognitive performance.

During its restoration of cell polarity after firing, a neurone uses ATP[23] and produces ADP[24] which needs to be converted back to ADP using glucose and oxygen. This places a high demand for oxygen on the respiratory and circulatory systems, requiring about 20% of the body's oxygen supply.

Even brief interruptions in oxygen delivery can lead to significant neuronal damage or death. When deprived of oxygen, for example by a cardiac arrest, consciousness is lost in about 15 seconds, and, at normal body temperature, irreversible cell damage occurs in about 4 minutes unless circulation is restored.

During its evolution the brain's high energy demands are a result of an evolutionary trade-off prioritizing survival advantages over energy efficiency. Improved survivability came about because the large, energy-intensive human brain favoured complex cognitive functions, such as problem-solving, language, and social interaction. These traits enhanced survival and reproductive success, outweighing the costs of maintaining a metabolically expensive organ.

The evolution of the human brain was closely tied to dietary changes, including the consumption of meat, fat, and cooked foods. These nutrient-dense diets provided the energy necessary to support a larger brain, however, this dependency makes humans more vulnerable to food shortages, emphasizing the

---

[23] Adenosine triphosphate.
[24] Adenosine diphosphate.

Part III: Inefficient or Wasteful Systems

evolutionary trade-off between brain development and dietary needs.

The brain's high demand for energy has also reduced the energy available for other bodily functions, such as physical strength or endurance. Compared to our closest cousin species, chimpanzees and gorillas, humans have a much smaller muscle mass making us puny by comparison – compensated for partly by cooperative groups working together.

This dependence on a constant high-energy supply has created several vulnerabilities that highlight its inefficiency:

> Stroke: The brain's reliance on a continuous blood supply means that blockages or haemorrhages in cerebral arteries can quickly cause severe damage.
>
> Hypoglycaemia: Drops in blood sugar levels impair brain function, leading to confusion, seizures, or unconsciousness.
>
> Neurodegenerative diseases: Conditions like Alzheimer's disease may be linked to the brain's high metabolic activity and the accumulation of metabolic byproducts.
>
> Developmental Trade-Offs: The energy demands of brain growth during childhood can delay other aspects of physical development, such as motor skills, resulting in an extended period of childhood dependence on parents.

Given the brain's importance to the survival and wellbeing of humans, an intelligently designed brain would prioritize efficiency and robustness in such a critical organ.

For example, an intelligently designed brain might store its own energy reserves or use multiple energy substrates more effectively. It could be optimized to achieve the same cognitive capabilities with reduced metabolic costs, and it could be designed to tolerate temporary interruptions in oxygen or glucose supply without significant damage.

The human brain's extraordinary energy demands testify to the trade-offs inherent in evolution. While natural selection has favoured cognitive capabilities that enhance survival and reproduction, these advantages come with significant costs and vulnerabilities. The brain's energy inefficiency and dependency on external resources reflect the iterative, patchwork nature of evolution, rather than the foresight and optimization expected of an intelligent designer.

Coning

Coning occurs when brain swelling pushes the brain into the brain stem where it emerges through the foramen magnum. This happens if a bleed occurs inside the brain tissue or between the brain and the skull, or because the brain is inflamed due to injury or infection, because there is no room for the increased volume so pressure in the cranium increases.

This places increased pressure on the blood vessels supplying the brain stem where the basic life-support systems such as blood pressure, cardiac output and respiratory rate and tidal volume controls are located. If pressure is not relieved surgically by, for example trepanning, this can be fatal.

The brain has evolved out of the primitive brain in the earliest chordates, where it provided automation for the basic life-support systems. Evolution continued through fish, amphibians and reptiles, and these basic autonomic functions have been retained in their original locations while the cerebral hemispheres and the more complex higher centres evolved in mammals.

Evolution has again, as always, prioritised function over longevity and durability.

Part III: Inefficient or Wasteful Systems

The problem of coning is due primarily to the lack of any pressure absorbing structures, and secondarily to the fact that life-critical centres are still located in the brain stem.

An intelligent designer could have relocated these centres and provided a pressure-relief system.

Concussion

The human brain's vulnerability to concussion and mental health disorders is a compelling area to examine in the evolution vs. intelligent design debate. These issues highlight significant structural and functional weaknesses in the brain that are difficult to reconcile with the idea of an omniscient designer. Instead, they align with the notion of a system shaped by evolutionary constraints and trade-offs.

A concussion is a type of traumatic brain injury (TBI) caused by a blow to the head or sudden acceleration/deceleration. Its effects, ranging from mild to severe, underscore the brain's structural fragility and reflect another evolutionary trade-off between function and structural durability and resilience.

The brain is a soft, jelly-like organ encased in a hard skull. Sudden movements or impacts can cause the brain to collide with the skull, leading to bruising (contusion) or tearing of nerve fibres. Evolution did not develop a robust cushioning system beyond the cerebrospinal fluid (CSF), which provides limited shock absorption.

By standing on two feet, not only are we less stable than a quadruped, but, if we trip, our head has further to travel before it hits the ground. We are also more liable to bang our head on a beam, low branch, or the low ceiling in a cave, but the brain's design has not fully adapted to mitigate this risk.

The large human skull has also evolved to house the large brain, making it more susceptible to injury from impacts.

Lastly, the brain is dependent on delicate, tightly-packed neuronal connections, making it especially susceptible to even minor trauma.

The short-term effects of concussion are headaches, confusion, memory loss, and dizziness. The longer-term effects, especially from repeated concussions, include chronic traumatic encephalopathy (CTE), a progressive condition characterized by mood disorders, cognitive decline, and neurodegeneration resembling dementia.

The brain also has zero capacity for regeneration, so damage is permanent, and any recovery is limited to retraining parts of the brain to take on new functions to replace those lost.

An intelligently designed brain would:

> Include better structural reinforcement to prevent damage during impacts (e.g., stronger skull cushioning or shock-absorbing tissues).

> Provide greater neural redundancy to ensure continued function despite injury.

> Incorporate regenerative mechanisms to repair damaged neurons.

The current design reflects a system that evolved under constraints, prioritizing cognitive ability over structural durability.

Mental health disorders

Mental health disorders, including depression, anxiety, schizophrenia, and bipolar disorder, represent another major weakness in the brain's design. These conditions are widespread, often debilitating, and poorly aligned with the concept of a perfect designer. Paradoxically, some mental disorders such as paranoid schizophrenia, have frequently been interpreted as voices from the locally-popular god (never one of the others), and the sufferer frequently believes they have a special God-given mission to save the world from disaster.

It's a measure of the cultural difference between the UK and America where the televangelist who tells his or her followers that they were, "Talking to God the other day and God said...", is treated with awe and reverence and people send them money. In the UK, people would wonder if their medication was due for review and bemoan the lack of beds in mental health-care facilities.

Mental health disorders themselves were once attributed to possession by evil spirits. Reading some of the 'prophesies' of the Old Testament prophets, to whom God allegedly spoke to convey essential messages and warning, it's difficult to avoid wondering if these were schizophrenic 'voices' they were hearing.

Mental health problems then were regarded as evidence of a god, not as the designer of perfection but as evidence of an interventionist god meddling in the affairs of Man and the existence of forces of evil.

They are, of course, evidence of neither. Instead, they are compelling evidence for evolution and serve to highlight the evolutionary compromises and trade-offs involved in a natural process that prioritises function over longer-term welfare.

The human brain evolved to support advanced cognitive and emotional processing. However, this complexity makes it vulnerable to dysfunction, for example, imbalances in neurotransmitters or neural circuitry can easily lead to mood disorders or psychosis.

Many mental health conditions may arise from an evolutionary mismatch between our brains, which evolved in small, tight-knit hunter-gatherer communities, and the stresses of modern life (e.g., social isolation, chronic stress, and information overload).

Certain traits linked to mental health conditions may have provided evolutionary advantages:

> Anxiety: Heightened vigilance could have improved survival in dangerous environments but becomes maladaptive in safe, modern contexts.
>
> Depression: Some researchers suggest depressive symptoms might have evolved to encourage introspection or social support during periods of adversity.
>
> Schizophrenia and Creativity: Some genetic factors associated with schizophrenia are linked to increased creativity, which might have been advantageous in small populations.
>
> Bipolar disorder: During the Hypermania phase of bipolar disorder, the person may have bursts of creativity and insight which are later over-compensated for by deep depression and lack of motivation.

Over 20% of the global population experiences a mental health disorder at some point in their life, indicating a widespread vulnerability. The maladaptive nature of these disorders is evidenced in impaired ability to function, form social bonds, or reproduce, representing a significant fitness cost.

If the brain had been intelligently designed, we would expect better emotional regulation systems to prevent disorders like depression or anxiety, neural mechanisms that are less prone to imbalance or dysfunction. The creativity and introspection that may have benefitted individual in small social groups could have been provided in a more stable way and would still provide benefits to the individuals in modern society.

The brain's susceptibility to mental health issues suggests a system that evolved through natural selection, with trade-offs that favoured short-term survival or reproductive advantages over long-term stability. It is thus compelling evidence against intelligent design.

The brain's vulnerability to concussion, coning, strokes, vulnerability to lack of oxygen and nutrients, and mental health problems provide strong evidence for evolution and challenges claims of intelligent design. These flaws serve to highlight the compromises inherent in an organ that evolved under constraints, prioritizing cognitive capacity over durability and emotional stability and having to make do with existing structures for new functions.

Evolutionarily, the brain reflects a patchwork solution to competing pressures, rather than the optimal creation of a designer.

Epilepsy

Epilepsy is another strong example that supports evidence for evolution over intelligent design. Epilepsy is characterized by abnormal electrical activity in the brain leading to seizures. It illustrates the vulnerabilities and inefficiencies in the human nervous system. These weaknesses are consistent with the evolutionary process of adapting pre-existing structures, rather than being indicative of a flawlessly crafted design.

Epilepsy is a neurological disorder that arises when the brain's electrical system becomes dysregulated. Seizures result from sudden, excessive bursts of electrical activity in the brain. The condition can vary widely in severity and frequency and may involve generalized seizures affecting both hemispheres of the brain or focal seizures, localized to one part of the brain. The two main types of epileptic seizures are the petit mal and grand mal seizures.

Petit mal seizures are periods of 'absence' lasting a few seconds to a few minutes in which the person is unaware of their surroundings but remains conscious.

Grand mal seizures have a tonic phase, a clonic phase and a post-ictal phase:

> During the tonic phase skeletal muscles contract violently so respiration ceases. This may last for seconds or minutes during which the sufferer may become cyanosed due to anoxia.
>
> During the clonic phase, skeletal muscles rhythmically contract and relax to give the rhythmic twitching more commonly associated with a fit. The sufferer is unconscious during the phase
>
> In the post-ictal phase, the sufferer has recovered motor control but may be disorientated and confused.

Both forms of epilepsy may be preceded by a 'aura' or auditory or visual hallucinations or feelings of *déjà vu*, which sufferers learn to recognise as a warning. Some focal seizures may manifest as a period of *déjà vu*.

Epileptic seizures are inconsistent with intelligent design because the human brain's reliance on precise electrical signalling makes it vulnerable to disruptions:

> The brain contains billions of interconnected neurons. While this complexity enables advanced cognitive abilities, it also makes the system prone to misfiring or feedback loops that

lead to seizures. There has been an evolutionary trade-off between cognitive abilities and the risk of malfunction.

Dysregulation of excitatory (e.g., glutamate) and inhibitory (e.g., GABA[25]) neurotransmitters can lead to neurotransmitter imbalance and abnormal electrical activity.

Brain malformations, scarring, or other physical issues can predispose individuals to seizures.

An intelligent design would include built-in safeguards to prevent electrical storms and redundant or fail-safe systems to counteract potential misfires.

Epilepsy reflects the constraints and compromises of evolution. In particular, the human brain evolved by exaptation of older neural systems, adding layers of complexity over simpler ancestral frameworks. This "retrofit" approach can lead to inefficiencies, such as susceptibility to electrical misfires.

The development of advanced cognitive functions, such as language and abstract reasoning, required more intricate neural circuitry. This complexity increases the risk of errors, including epilepsy.

Achieving a function by adding layers of complexity to retrofit existing structure for novel functions is typical of the make-do evolutionary process, constrained as it is to work with whatever is to hand and accept whatever works. It is the antithesis of intelligent design which could build from scratch.

---

[25] Gamma-aminobutyric acid (GABA) is the most common inhibitory neurotransmitter in your central nervous system. GABA lessens the ability of a nerve cell to receive, create or send chemical messages to other nerve cells. GABA is known for producing a calming effect. It's thought to play a major role in controlling anxiety, stress and fear.

Epilepsy often has genetic or environmental triggers, further illustrating the imperfections of the brain's design:

> Many cases of epilepsy are linked to mutations in genes that regulate ion channels or neural activity. These mutations arise as part of the random genetic variations on which natural selection acts, not as part of a perfect design.

> Brain injuries, infections, or stress can trigger epilepsy. The lack of resilience to such common occurrences suggests an imperfectly adapted system.

Certain aspects of epilepsy reflect evolutionary trade-offs:

> The brain's excitability is critical for functions like learning, memory, and creativity. However, this excitability also increases the risk of seizures.

> Hyperexcitable brains may have been advantageous in some ancestral environments, fostering rapid problem-solving or heightened responses to danger.

Epilepsy's debilitating effects and prevalence pose a challenge to the concept of a perfect designer:

> Seizures impair essential brain functions, such as consciousness, memory, and motor control.

> Seizures can occur suddenly and unpredictably, endangering the individual in daily life.

> Untreated epilepsy can significantly reduce survival and reproductive success, suggesting a costly flaw in brain design.

An intelligently designed brain would include:

> Enhanced mechanisms to prevent electrical misfires.

Part III: Inefficient or Wasteful Systems

Adaptations to minimize the impact of genetic mutations or environmental damage, even if the designer was powerless to prevent them.

Greater robustness against excitatory-inhibitory imbalances.

Epilepsy is a striking example of the brain's vulnerabilities and inefficiencies, consistent with evolution rather than intelligent design. Its existence highlights the imperfect, incremental nature of evolutionary processes, which build upon pre-existing structures without the foresight or optimization expected of an intelligent creator. The condition underscores the trade-offs inherent in the evolution of a complex organ like the human brain

Incidentally, some forms of temporal lobe epilepsy, i.e. seizures originating from a focus in a temporal lobe, are associated with hearing voices, hallucinations and a feeling of presence, even a spiritual or religious experience. For an almost perfect description of a temporal lobe seizure one could do worse than read the account of the 'conversion' of Saul of Tarsus in the New Testament of the Christian Bible, when he reported 'seeing the light' (Acts 9:3-8)

The Immune System's Paradox

The penultimate part of this section on inefficient and wasteful systems is the evolution of the immune system and why it has given rise to problems such as autoimmune conditions and allergies while not providing effective protection, especially against organisms which have evolved to evade it.

Autoimmune conditions and allergies provide strong evidence for evolution over intelligent design by highlighting the imperfections and trade-offs inherent in the immune system. These disorders demonstrate how a system shaped by natural selection can prioritize short-term survival at the expense of long-term health, leading to vulnerabilities that are difficult to reconcile with the concept of a perfect designer.

Autoimmune diseases occur when the immune system mistakenly attacks the body's own tissues. Examples include rheumatoid arthritis, lupus, multiple sclerosis, and type 1 diabetes.

The immune system must strike a delicate balance: it must be reactive enough to fight infections but tolerant enough to avoid attacking the body's own cells. Evolution has shaped this balance, but it is imperfect. A hyperactive immune system, while better at combating infections, increases the risk of autoimmune diseases.

In September 2024, a Japanese research team at the Tokyo Medical and Dental University showed that a common oral bacterium, *Aggregatibacter actinomycetemcomitans*, is closely associated with the onset and worsening of the autoimmune condition, rheumatoid arthritis (22).

In modern, cleaner environments, according to the 'hygiene hypothesis' the immune system may lack the natural challenges (like infections or parasites) that it evolved to handle, leading to overactivity and autoimmunity.

This mismatch between our evolved immune system and modern environments underscores its suboptimal design.

Certain genetic variants that boost immune activity against pathogens may also predispose individuals to autoimmunity. For example, the HLA gene family[26], crucial for immune recognition, is often implicated in autoimmune diseases.

---

[26] The human leukocyte antigen (HLA) gene family is a group of genes on chromosome 6 that produce proteins that help the immune system recognize foreign invaders

Part III: Inefficient or Wasteful Systems

Autoimmune conditions can cause chronic pain, disability, and even death, with no apparent evolutionary benefit once an individual has passed their reproductive years. These diseases are particularly incompatible with the notion of an intelligent designer who would presumably avoid such self-destructive flaws, unless autoimmune diseases and the suffering they cause is part of the design.

Allergies occur when the immune system overreacts to otherwise harmless substances, such as pollen, food, dust mites, nuts, or the sting of certain normally relatively harmless insects. Severe allergic reactions, such as anaphylaxis, can be life-threatening.

Allergies may represent an overzealous defensive response to perceived threats. Evolution likely favoured a "better safe than sorry" approach, where false positives (reacting to harmless substances) were less costly than false negatives (failing to react to real threats, like venomous bites).

For example, the immune system's IgE pathway, which mediates allergic responses, is thought to have evolved to defend against parasites, particularly helminths[27]. In modern environments with fewer parasites, this system may misfire, leading to allergies.

Again, according to the hygiene hypothesis, reduced exposure to pathogens and parasites in childhood may cause the immune system to become "bored," redirecting its activity toward harmless antigens, resulting in allergies. This hypothesis emphasizes the mismatch between ancestral and modern environments.

Some researchers propose that allergic responses, such as sneezing or itching, may have evolved as a way to expel

---

[27] Round or flat worms such as liver fluke, tapeworms and intestinal nematode parasites.

pathogens or harmful substances. However, in modern contexts, this response is often excessive and maladaptive.

A severe allergic reaction cab produce anaphylaxis, a severe and potentially fatal allergic reaction, demonstrates a catastrophic failure of immune system regulation.

The idea that an intelligent designer would create a system where exposure to harmless substances like peanuts or bee stings could result in death is utterly implausible and refutes any notion of intelligent design.

What we have in the flaws in the immune system that cause allergies and autoimmune diseases, is a classic example of evolutionary trade-offs, not perfect design, by any stretch of the imagination or intellectual gymnastics.

These issues likely stem from the immune system being shaped by natural selection to maximize immediate survival in pathogen-rich environments, even at the expense of long-term health. Short-terms survival was prioritized over longevity, as it inevitably does.

The prevalence of autoimmune diseases and allergies in modern societies highlights how evolution cannot predict or prepare for rapid environmental changes. Many genetic variants associated with autoimmunity or allergies may have been beneficial in past environments but are maladaptive today.

These problems refute the notion of an intelligently designed immune system which, by any definition of the word 'intelligent', would not attack the body or overreact to harmless stimuli.

An intelligently designed system would also adapt seamlessly to changing environments, unlike the immune system we observe, which often struggles to cope with modern conditions.

Part III: Inefficient or Wasteful Systems

Autoimmune diseases and allergies cause significant suffering and disability, with no clear benefit. This is inconsistent with the idea of a benevolent designer.

The prevalence of autoimmune diseases and allergies underscores the immune system's flaws and inefficiencies, which are better explained by evolution than by intelligent design. These conditions highlight the compromises and trade-offs inherent in a system shaped by natural selection, providing compelling evidence for evolution's explanatory power.

The Wasteful Reproductive Strategy

The final topic in this section looks at the human reproductive strategy in terms of its resource and energy efficiency, or rather the glaring lack of these.

As I showed in my book, *The Unintelligent Designer: Refuting the Intelligent Design Hoax* (10) prolific waste and unnecessary complexity are hallmarks of evolution not intelligent design, the hallmarks of which are maximal simplicity and minimal waste. The waste inherent in the human (and other mammalian) reproductive systems is consistent with an evolutionary process and utterly inconsistent with an intelligent design process.

We've already seen how the male and female reproductive systems are maladapted for modern living, being the result of evolutionary trade-offs associated with bipedalism and an upright posture where evolution has prioritized for short-term reproductive advantage over longevity and the pain and suffering associated with pregnancy and childbirth. As such it is a compelling argument against intelligent design by an omnibenevolent designer.

The wasteful reproductive strategy observed in humans and other species is a compelling argument against intelligent design and for evolution. Reproductive processes, from gamete production to pregnancy, are characterized by inefficiencies, high energy costs, and significant losses. These characteristics reflect the constraints and compromises inherent in evolution, as opposed to the foresight and optimization one might expect from an intelligent designer.

Creationists often claim, for example that each conception is a 'gift from God' with every child born being exactly the child that God intended. This begs the question then, why hundreds of millions of sperms are directed towards the egg, with the winner being the one that gets there first. If a designer god had intended just that conception with just that DNA, why did it not arrange for a single sperm with the right DNA to fertilise the egg? What were all the others for?

Of course, that creationist belief is a religious opinion having no basis in biology or reproductive physiology. The hundreds of millions of wasted sperms, each with half the adult human genome, represent a colossal waste of energy and nutrients because evolution has optimised on reproductive success over resource efficiency. The strategy reflects evolutionary pressure to maximize the likelihood of fertilization in competitive environments, not efficiency.

It has been said that humans are mostly monogamous, most of the time. This was probably true in our evolutionary ancestors, so, with females capable of mating with multiple males, it pays to produce the most sperm because that increases the probability of fertilising the egg and fathering a child.

There would have been strong evolutionary pressure to produce more and more sperm because the male who produced the most sperms per ejaculation would tend to have the most offspring to inherit the genes for a large number of sperm.

Part III: Inefficient or Wasteful Systems

Females are born with approximately 1–2 million oocytes, but only about 400 will be ovulated over a lifetime. The rest undergo atresia (degeneration).

This wasteful process is a byproduct of an evolutionary strategy where more eggs were initially produced than could ever be used, ensuring that only the most viable are selected for ovulation.

Even with large numbers of sperm, many never reach the egg due to anatomical barriers, hostile environments in the female reproductive tract, or poor sperm quality. This inefficiency underscores an evolutionary "trial and error" approach rather than a precise, intelligently guided system.

Studies suggest that up to 60% of fertilized eggs fail to implant in the uterus or are spontaneously aborted shortly after implantation. Most of these losses are due to chromosomal abnormalities or other developmental issues. An intelligent designer might have created a system that ensures only viable embryos are fertilized, avoiding the wasteful loss of fertilized eggs.

Miscarriages occur in approximately 10–20% of known pregnancies, often due to genetic or developmental issues. Stillbirths, though less common, still represent a tragic inefficiency in the reproductive process.

These losses, while partially explicable through natural selection (to prevent the birth of offspring with severe defects), represent a stark inefficiency inconsistent with the notion of intelligent design.

As I explained earlier, the human menstrual cycle is energetically costly, involving the regular shedding and regeneration of the endometrium. In contrast, some mammals (e.g., dogs) reabsorb the endometrial lining, conserving energy.

The Body of Evidence

I discussed the complications inherent in human pregnancy and how it is incompatible with notions of intelligent design earlier, but to recap:

> Pregnancy places immense metabolic demands on the mother, requiring increased caloric intake, oxygen supply, and resource allocation.

> The narrow human birth canal (a consequence of bipedalism and large brain evolution) makes childbirth risky and painful, often requiring medical intervention.

> An intelligent design would presumably minimize the risks and costs associated with reproduction, rather than leaving such a narrow margin for survival.

The wasteful reproductive strategy reflects the evolutionary need to ensure species survival in environments with high levels of unpredictability, including predation, disease, and resource scarcity. Overproduction of gametes and embryos is a "numbers game," increasing the likelihood that some offspring will survive despite high attrition rates.

Evolution relies on genetic diversity to drive adaptation. The production of excess gametes, combined with natural selection, ensures that only the most fit individuals contribute to the gene pool. The trade-off is the prioritization of survival and reproductive success in the short term, even if the process is wasteful or causes long-term harm.

The inefficiencies and risks of human reproduction runs counter to what an intelligent designer would produce:

> A perfect designer would not create a system that wastes billions of sperm and millions of eggs.

Part III: Inefficient or Wasteful Systems

An intelligent system would minimize the loss of embryos and miscarriages, ensuring efficient and successful reproduction.

The dangerous and painful nature of human childbirth suggests a poorly optimized design.

Pregnancy and menstruation are high-cost processes that a benevolent designer could have streamlined.

The wasteful reproductive strategy in humans is a striking example of the evolutionary process in action. Its inefficiencies, high failure rates, and substantial costs make little sense in the context of intelligent design but are consistent with the trial-and-error nature of evolution. This reproductive system reflects the compromises and imperfections that arise from natural selection working with existing structures, not the foresight or perfection of a deliberate creator.

The Body of Evidence

Part IV: Evolution Explains it All.

The Body of Evidence

Part IV: Evolution Explains it All

Vestigial Organs, Atavisms and Exaptation

Vestigial organs and atavistic traits in the human body provide some of the most compelling evidence for evolution and stand in stark contrast to the concept of intelligent design. These features are remnants of our evolutionary past, serving little or no current function but often reflecting functions that were important to ancestral species.

The only reason an intelligent designer would have included vestigial organs, and atavistic traits would be to make it look like the human body had evolved from ancestral species in which these structures were important.

To believe an omnibenevolent god would deliberately deceive us by forging the Universe to make it look old and life on Earth to make it look like it had evolved as some sort of test, is to believe it didn't want us to use our brain to discover the truth. It is a strange view of such a god that it lied in all the physical evidence it created, and you know this because it once told the truth and inspired a book to relate it in.

A single book with a dubious history, full of unlikely tales, contradictions and factual inaccuracies is more dependable than the evidence a creator god created!

Vestiges of ancient ancestry

The following are some examples of vestigial organs in humans:

Coccyx (Tailbone):

> The coccyx is the small triangle of 3-5 small bones at the base of the spine, a remnant of the tails possessed by our primate ancestors. In ancestral species with tails, the coccyx served as a critical attachment point for muscles involved in tail movement.

The coccyx persists as a structural leftover and can cause significant pain (coccydynia) when injured, serving no practical function but posing a liability.

Wisdom Teeth

As I mentioned in Part I in relation to the digestive tract, of which teeth form a part, in early humans and hominins, larger jaws accommodated more teeth, and these molars were essential for grinding fibrous plant material. As human diets changed and jaws became smaller, wisdom teeth became redundant but were not eliminated by evolution.

Impacted wisdom teeth frequently require surgical removal, causing pain, infection, and complications, suggesting an evolutionary mismatch rather than intelligent design.

Plica Semilunaris (Third Eyelid)

The small fold of tissue in the inner corner of the eye, homologous to the nictitating membrane in some reptiles and birds, which keeps the eyeball clean and protected without being closed. In a peregrine falcon, for example, a 200 mph stoop onto prey carries a high risk of damage to the cornea from dust and flying insects and damaged eyes would be fatal for a raptor that depends on them for hunting, so a nictitating membrane is essential.

Ancestors of mammals likely had a functional third eyelid for protecting and moistening the eye. It has been retained in some species such as domestic cats but in humans, it has become a vestigial structure with no known function. Its presence is neutral but serves as a reminder of unused evolutionary baggage.

Part IV: Evolution Explains it All

Erector Pili Muscles (Goosebumps)

The erector pili muscles, also known as arrector pili muscles, are small muscles attached to hair follicles that contract to produce goosebumps and make the hair stand up. Each muscle is supplied by a sympathetic nerve that releases noradrenaline (norepinephrine) to cause it to contract. This nervous control is activated as part of the 'fight or flight' response and in response to a lowering of the skin surface temperature.

In furry ancestors, these muscles caused hair to stand on end, providing insulation and making them appear larger to predators. In chimpanzees and gorillas, the hair on the back shoulders and arms stands out during aggression or in response to a threat, but in humans, this mechanism is largely ineffective due to reduced body hair.

The energy and biological resources used to maintain these muscles and their nerve supply could be better allocated elsewhere if the system were intelligently designed.

Auricular Muscles (Ear Movement)

Some of us have muscles that can be used to wiggle our ears voluntarily and make them stand out slightly, automatically when startled. These auricular muscles allowed ancestral mammals to swivel their ears to detect sounds more effectively. Humans retain these muscles but have largely lost the ability to use them for auditory advantage. Although they can still be used to amuse grandchildren!

The persistence of these muscles without function highlights the inefficiency of evolutionary leftovers and is the antithesis of intelligent design.

The Body of Evidence

The Preauricular Pit.

The preauricular pit (or preauricular sinus) is a small congenital opening or pit located near the front of the ear, where the ear's helix meets the face. While generally harmless, it is a vestigial feature whose evolutionary origins can be traced back to the development of the ear and its ancestral structures.

During embryogenesis, the external ear (auricle) develops from six small swellings of tissue called auricular hillocks, which arise from the first and second pharyngeal arches. These hillocks eventually fuse to form the outer ear. Errors in this complex fusion process can lead to the formation of preauricular pits. The pits are thus developmental anomalies, reflecting the complex evolutionary history of ear morphology.

The structures involved in ear development (pharyngeal arches) are homologous to the gill arches in fish, a reminder of our aquatic ancestors. Preauricular pits may represent a vestigial feature related to the branchial clefts in fish. While speculative, this connection highlights the evolutionary transition from gill-breathing organisms to land-dwelling vertebrates.

Early mammalian ancestors probably had simpler external ear structures, and features like preauricular pits may represent remnants of transitional stages in the evolution of complex mammalian ears. Mammalian ears evolved to enhance directional hearing, a significant adaptation for nocturnal survival in early mammals.

In most cases, preauricular pits are benign and do not serve any functional purpose, supporting the idea that they are vestigial remnants rather than adaptive traits, however they can become infected or form cysts, refuting the notion of intelligent design in a structure that has no function but poses risk, no matter how slight.

Preauricular pits are relatively rare, occurring in about 0.1–10% of the population, with higher prevalence in certain regions (e.g., up to 10% in parts of Africa and Asia compared to 0.1–0.9% in the U.S.). This distribution and frequency may reflect genetic drift or neutral evolution rather than selection.

So, why have they been retained at all?

Preauricular pits are generally harmless and do not affect survival or reproduction, so natural selection does not eliminate them. The complexity of ear development means that minor variations, like preauricular pits, may occur as a byproduct of the intricate process, without significant evolutionary consequences.

The preauricular pit is an example of a vestigial feature that reflects our evolutionary past, particularly the transition from aquatic to terrestrial life and the gradual development of mammalian ear structures. While it serves no known purpose today, its persistence underscores the non-intelligent, tinkering nature of evolution—where ancestral remnants remain, not because they are useful, but because they are not harmful enough to be eliminated.

Atavistic traits

After those examples of vestigial structures that serve no useful purpose and for which it is impossible to make a case for their creation by an intelligent designer, I'll turn now to atavistic traits in humans. Atavisms are traits that reappear in an organism after being absent in several generations, providing clear evidence of ancestral traits encoded in our DNA.

The Body of Evidence

There are two main causes of atavism: either the deactivated gene which used to control their development in an ancestor, and which was deactivated as part of the evolutionary process, becomes reactivated, or a gene which evolved to suppress the genes controlling its development, are deactivated or absent.

Tail Formation

Some humans are born with a small, non-functional tail, which is surgically removed shortly after birth. This rare occurrence represents a reactivation of genes for tail development, a feature prominent in our distant mammalian ancestors.

The reappearance of an outdated trait serves no purpose and underscores the imperfect retention of ancestral genes.

Extra Nipples (Polythelia)

Some people are born with extra nipples along the "milk line," a feature common in mammals that bear multiple offspring.

Extra nipples are a vestige of our mammalian ancestry, where multiple nipples were necessary for feeding large litters. In humans, extra nipples are functionless, serving as a reminder of evolutionary history rather than intentional design.

Dense Body Hair (Hypertrichosis)

Excessive hair growth resembling the fur of ancestral species.

Hypertrichosis is thought to occur due to the reactivation of dormant genetic pathways that were active in our hairier ancestors.

The sporadic reactivation of an obsolete trait serves no purpose and contradicts the idea of purposeful design.

Part IV: Evolution Explains it All

Vestigial organs and atavistic traits are direct, observable evidence of evolution's trial-and-error process. These remnants of our evolutionary history highlight inefficiencies, redundancies, and imperfections in human anatomy. Such features are incompatible with the idea of an intelligent designer who would presumably create optimized, purposeful structures. Instead, they align with the predictions of evolution, where natural selection repurposes existing structures for new functions but leaves traces of the past behind.

Exaptation of redundant DNA.

There are many examples of redundant genes, some of which have been exapted for entirely different functions, illustrating how evolution builds on what is available, in this case, chunks of DNA that became redundant coding for novel enzymes or structures. This, incidentally, gives the lie to frequent creationist assertions that the evolution of new genetic information is impossible because it would violate the Laws of Thermodynamics akin to creating new energy.

It is nothing of the sort, of course. Simply substituting one amino acid for another in a protein doesn't create any new matter or energy, it simply changes the meaning of the information. The creationist assertion is akin to claiming a typist couldn't make a mistake and change 'make' into 'mate' because the Laws of Thermodynamics forbid it.

In the human genome, several examples exist where redundant or non-essential genes have been exapted for new purposes. These cases highlight how evolution works by repurposing pre-existing structures rather than creating entirely new features from scratch.

Genes from Endogenous Retroviruses (ERVs)

ERVs are remnants of ancient viral infections that inserted their genetic material into the human genome. These sequences were originally parasitic and served no beneficial function.

Some ERV genes have been co-opted for essential functions:

This gene, derived from an ERV, is crucial for placental development, specifically in forming the syncytiotrophoblast layer that facilitates nutrient exchange between mother and foetus. Without this gene, the evolution of placental mammals would have been unlikely.

Other ERV sequences regulate nearby genes, including those involved in immune responses.

Originally redundant or harmful, these viral genes have become indispensable for human reproduction and immunity.

Pseudogenes Acting as Regulatory Elements

Pseudogenes are non-functional remnants of genes that have lost their ability to code for proteins. Some pseudogenes regulate the expression of functional genes by acting as decoys or sponges for regulatory RNAs (e.g., microRNAs).

For example:

PTENP1, a pseudogene of the tumour suppressor gene PTEN, helps regulate PTEN expression by sequestering microRNAs that would otherwise suppress PTEN activity.

What was once a redundant or "junk" gene now plays an active role in gene regulation and cancer prevention.

Part IV: Evolution Explains it All

Gene Duplication and Divergence

Duplication of existing genes creates redundant copies that may initially have no new function. In humans, many genes have been exapted for novel functions after duplication:

Amylase (AMY1): Duplication of the amylase gene in humans has increased the production of salivary amylase, aiding in starch digestion. This adaptation is particularly evident in populations with high-starch diets.

Globins: The duplication of ancestral globin genes led to the evolution of haemoglobin subunits (alpha, beta, gamma), which are specialized for oxygen transport in different developmental stages (e.g., foetal haemoglobin for oxygen transfer from mother to foetus).

Gene duplications provide raw material for evolutionary 'experimentation', leading to innovations in metabolism and oxygen transport.

Transposable Elements as Regulatory Elements

Transposable elements (TEs) are "jumping genes" that move around the genome, often considered selfish DNA.

In humans, TEs have been repurposed as regulatory sequences that control gene expression. For example, Alu elements (a type of TE) influence splicing, transcription, and gene regulation.

The MER41 family of TEs has been co-opted to regulate immune response genes, including those involved in interferon activation.

These redundant or parasitic sequences have become integral to human gene regulation and immune function.

FOXP2 and Speech

The FOXP2 gene, involved in speech and language, evolved from a more general-purpose regulatory gene. Specific mutations in FOXP2, unique to humans, have enhanced its role in neural circuitry related to vocal learning and articulation.

FOXP2's evolution showcases how a gene can be modified and specialized for a novel function critical to human communication.

Human-Specific Enhancers

Enhancers are regulatory DNA sequences that control gene expression. Many enhancers in humans derive from pre-existing sequences, including transposable elements or non-functional DNA. Some enhancers unique to humans regulate genes involved in brain development, such as those driving the growth of the neocortex.

For example, HARE5, an enhancer linked to brain size, has been shown to drive increased neural growth in human-specific patterns.

Redundant or dormant sequences were co-opted to drive critical innovations in human brain evolution.

Vitamin C Synthesis Gene (GULO)

Humans have a pseudogene (GULO) for the enzyme necessary for vitamin C synthesis, a capability lost in primates due to dietary sufficiency in fruits. While the GULO gene itself is non-functional, related pathways for oxidative stress management have been exapted or enhanced, reflecting adaptations to alternative metabolic roles.

The loss of vitamin C synthesis and subsequent reliance on diet exemplifies evolutionary trade-offs, where redundant pathways can lead to new dependencies.

Incidentally, the loss of the ability to synthesise vitamin C, which occurred in a remote simian fructivore ancestor, is common to all members of the branch of the mammalian family tree that includes all the great apes. It was caused by breaking the gene for the fourth enzyme in a four-step metabolic pathway by a simple frame shift. The other three genes remain active and still produce the first three enzymes.

Where is the intelligence in providing the first three genes and then breaking the fourth, when it would have been simpler to just not have all four genes?

The human genome is filled with examples of exaptation, where once-redundant or "useless" genetic elements have been repurposed for critical biological roles. These examples demonstrate the evolutionary process of tinkering—recycling existing material to create new functions—contrasting starkly with the notion of a perfect, intelligent design.

Path Dependence in Evolution

Path dependence in evolution refers to how new traits and structures evolve by modifying pre-existing ones, constrained by historical and developmental pathways. This principle explains why evolutionary solutions are often suboptimal, as evolution works with what is already present rather than starting from scratch. In humans, path dependence manifests in numerous anatomical and physiological features, providing strong evidence for evolution and challenging the notion of intelligent design.

Some examples are:

The Body of Evidence

The Pharyngeal Arches and Craniofacial Development

The pharyngeal arches, initially gill-support structures in fish, were repurposed during vertebrate evolution. In humans these arches give rise to multiple head and neck structures, including the jaw, middle ear bones, and throat.

The recurrent laryngeal nerve, which loops under the aortic arch to reach the larynx, reflects its evolutionary origin in fish, where the heart is closer to the gills. In humans, this circuitous route creates inefficiency and vulnerability. It becomes absurd in the giraffe!

A direct pathway for the nerve would be far more efficient. The current structure only makes sense as a legacy of fish anatomy, modified incrementally over millions of years by a utilitarian process acting without a plan

The Human Spine

The human spine evolved from the horizontal vertebral column of quadrupeds. The transition to bipedalism required modifications for upright posture, but the spine retains many features of its quadrupedal ancestry.

The human spine is prone to problems like herniated discs, scoliosis, and lower back pain due to the stresses of vertical load-bearing.

An intelligent design would likely involve a spine optimized for bipedal locomotion, rather than one adapted from a quadrupedal structure prone to chronic pain and injury.

The Human Eye

The vertebrate eye evolved from simpler light-sensitive structures. The retina is inverted, with photoreceptor cells buried beneath layers of neurons and blood vessels.

This design creates a blind spot where the optic nerve exits the eye, and light must pass through the retinal layers before reaching photoreceptors.

An intelligent design would avoid the blind spot and place photoreceptors directly facing incoming light, as seen in the eyes of cephalopods like octopuses.

As I mentioned earlier, there are many superior designs of the eye, even in vertebrates, so the mediocre human eye belies the notion the humans are at the pinnacle of an intelligent creator's creation.

I give several more examples of where humans get a raw deal as the design of an intelligent designer in *The Unintelligent Designer: Refuting the Intelligent Design Hoax* (10).

The Human Pelvis and Childbirth

I probably said enough about the human pelvis and the problems its remodelling for bipedalism and an upright posture has cause in Part I.

This trade-off reflects the constraints of evolving a bipedal pelvis while also accommodating large-brained infants.

An optimal design would avoid this dangerous trade-off, ensuring both efficient locomotion and safe childbirth.

The Vas Deferens in Males

The vas deferens develops from the mesonephric duct, an embryonic structure linked to primitive excretory systems. Consequently, it takes a convoluted route, looping around the ureter before reaching the urethra.

This inefficiency reflects the duct's evolutionary origin and repurposing for sperm transport.

A direct, efficient pathway for sperm transport would be expected in an intelligently designed system.

Path dependence in human anatomy demonstrates how evolution repurposes and modifies pre-existing structures, constrained by historical and developmental factors. This results in suboptimal designs, inefficiencies, and vulnerabilities that make sense only in the context of evolutionary history.

Such evidence starkly contrasts with the notion of an all-knowing, intelligent designer who could create optimal, flawless systems from scratch. Instead, it highlights evolution's trial-and-error process, where past constraints shape present outcomes.

Conclusion

The human body is a patchwork of ingenious adaptations, redundant features, and glaring inefficiencies. Far from suggesting a perfect, intelligent design, it instead speaks eloquently to the history of evolution: a process constrained by path dependence and driven by natural selection, mutation, and genetic drift. By examining our anatomy and physiology, we uncover compelling evidence for evolution, while simultaneously challenging the notion of intelligent design.

To recap the main points:

The skeletal system provides a record of the transition from quadrupedalism to bipedalism, while simultaneously remodelling the head and face to accommodate a large brain and rotate the face through 90 degrees relative to the spine. This process involved the inevitable trade-offs and compromises inherent in the evolutionary process.

This resulted in suboptimal solutions to the problems caused by the evolutionary process's inevitable prioritisation of short-term advantage over longevity and efficiency with the layers of additional, error-prone complexity that characterise unplanned, unintelligent evolution and distinguish it from intelligent design which would be less complex therefore less error-prone and more energy and resource efficient.

The human spine evolved from the horizontal vertebral column of quadrupedal ancestors, modified for bipedalism. This transition was necessary for upright posture but introduced significant drawbacks. The S-shaped curve of the spine, while aiding balance, predisposes humans to conditions like herniated discs, scoliosis, and chronic lower back pain. These vulnerabilities highlight the compromises inherent in evolutionary adaptations, where pre-existing structures are repurposed rather than redesigned.

The human pelvis represents another compromise. Adapted for bipedal locomotion, it has a narrowed birth canal, making childbirth difficult and dangerous. Large-brained infants exacerbate this problem, resulting in high maternal and neonatal mortality rates in the absence of medical intervention. This trade-off between locomotion and childbirth efficiency is far from optimal, underscoring evolution's tinkering rather than a flawless design.

The coccyx is a remnant of a tail from our primate ancestors. It serves minimal functional purpose, providing support for pelvic organs and muscles. Rare cases of congenital tails in humans further affirm its vestigial nature. Similarly, wisdom teeth—once useful for chewing tough plant material—frequently cause overcrowding and pain in modern jaws, reflecting changes in diet and jaw structure over evolutionary time.

Evolution has failed to mitigate the additional problems an upright posture and a remodelled pelvis has caused for the pelvic floor, which now needs to bear the weight of the organs of the lower abdomen, increasing the risk of prolapsed uterus and stress incontinence in women and inguinal hernias in men.

The Digestive System carries a similar records of evolutionary compromises and exaptations, for example, the vermiform appendix, once thought to be entirely redundant, is now understood to play a minor role in immune function. However, its primary significance lies in its evolutionary history. As a vestige of a larger cecum used by herbivorous ancestors for digesting cellulose, it is prone to inflammation and life-threatening appendicitis. This susceptibility underscores its inefficiency and redundancy.

Conclusion

In addition to problems caused by postural changes and the way gravity is affecting organs that evolved in a different orientation are weaknesses and vulnerability stemming from a failure to anticipate and adapt to changes in diet and the stresses of modern life.

For instance, diverticula, small pouches that can form in the colon, and their associated inflammation, diverticulitis, highlight another flaw in the human digestive system reflecting structural weaknesses that arise from evolutionary compromises in gut design. And both ulcerative colitis and IBS exemplify how the human digestive system remains vulnerable to dysfunction. These disorders may stem from the evolutionary trade-offs associated with immune function and stress response, as well as the challenges posed by modern diets and lifestyles.

Remnants of former optimizations for an earlier lifestyle can be seen in, for example an exaggerated response to possible foodborne toxins during pregnancy which manifests as hyperemesis gravidarum. This also illustrates the imprecise nature of a near-enough-is-good-enough evolutionary process which fails to discriminate between serious toxins and harmless ones, producing a response which can but both mother and baby as risk – the opposite of what an intelligently designed defence mechanism would produce.

In the reproductive systems of both men and women we have seen the failure to produce optimal solutions, in, for example the failure to fully compensate for the additional demands of the large brain the foetus is growing, which has resulted in a deep penetration of the placenta into the endometrium. This has increased the risk of the potentially fatal placenta previa and pre-eclampsia due to the competing nutritional needs of mother and baby and the tendency for the baby to be more like a parasite in the body of a host, rather than the symbiotic relationship an intelligent designer could have devised.

Menstruation, which is thought to have evolved to protect the uterus from pathogens, was optimized for an environment in which pathogens were more of a risk than with modern hygiene and possibly because dissociating sexual intercourse from a purely procreational function to a recreational, social-bonding role increases the risk of infections. However, it is a resource-intensive and potentially debilitating process. Its persistence underscores evolutionary path dependence rather than an optimized reproductive strategy.

Locating the testes outside the abdominal cavity because spermatogenesis evolved in cold-blooded vertebrates creates the problem of inguinal hernia and testicular torsion while making the testes vulnerable to injury. In mammals such as elephants and manatees, this problem has been overcome so the testes remain in the abdominal cavity, so there appears to be no reason an intelligent designer could not have provides a more efficient, less risky solution.

More examples of suboptimal design and evolutionary compromises can be found in the circulatory and respiratory systems, not the least of which is the risk of choking because the airway crosses the food path in the pharynx but the lengthened pharynx optimised for complex speech has placed the top of the larynx further away from the epiglottis which should close it off during swallowing. Evolution has optimised on speech at the expense of safety.

Blood clotting is dependent on an evolved, multistep process in which complexity has added speed and efficiency but at the risk of thrombosis due to a lack of the ability of the triggering mechanism to discriminate between harmless and harmful conditions. An intelligent designer could have provided a

Conclusion

simpler, more discriminating mechanism for preventing blood loss from injury.

The coronary circulation depends on maintaining an adequate blood pressure during the heart's resting phase and the 'end veins' are prone to thromboses forming on plaque build-up in these arteries. A single artery supplies a major part of the cardiac muscle with no redundancy as a safety net for a critical system.

Bipedalism and an upright posture have increased the risk of varicose veins in the legs because the veins are not adapted to the additional burden of needing to return the blood against gravity. The evolutionary constraints and suboptimal compromises inevitable in an evolutionary process have produced a debilitating condition which occurs too late in life to have any selection pressure to mitigate the effects – illustrating how the evolutionary process optimizes on short-term gain over long-term welfare, pain and suffering, in contrast to what one might expect of an intelligent, omnibenevolent designer.

The human brain is an extremely complex organ which shows evidence of its evolutionary origins in its vulnerabilities to oxygen deprivation, injury, and lack of robust solutions to misfiring and the resulting epileptic seizures. The various mental health problems may be the result of traits evolved in smaller, hunter-gatherer social groups, but which have failed to evolve to compensate for the stresses and strains of modern living.

It is also energy intensive consuming about 20% of the human body's energy rest. This has provided remarkable cognitive and communicative abilities but has made the body vulnerable to starvation and metabolic stress.

The Body of Evidence

The brain is highly dependent on a supply of oxygen and glucose in the blood because it has no capacity to store either. This leaves it at substantial risk if deprived of either, with consciousness being lost within seconds of a cardiac arrest and irreversible brain damage occurring within minutes. Cognitive ability has been prioritised over nutrient and oxygen storage capacity and the fatal results of this deficit in design carry low selection pressure because they tend to occur after reproduction has ceased. The short-term has inevitably been prioritised over longevity, contrary to what an intelligent design process would produce.

The brain's vulnerability to concussion and the prevalence of mental health disorders underscores its evolutionary limitations. While these traits reflect trade-offs for advanced cognitive and social functions, they challenge the notion of intelligent design.

Further problems have been caused by the immune system, which often fails to protect us from the pathogens which are ahead in their arms race with our immune system and can attack us from a number of different directions despite better hygiene and safer sources of food.

Our immune system evolved in less hygienic times and the evolutionary process evolved a hypersensitivity to pathogens but a lack of the sensitivity needed to distinguish between the body's own tissues and those of invaders, so there is a tendency to produce autoimmune conditions which can be life-limiting or rapidly fatal in the case of an anaphylactic response to a harmless stimulus such as nut allergy.

Conditions such as lupus, rheumatoid arthritis and Type I diabetes are autoimmune conditions that reflect the suboptimal immune system, that an intelligent designer could have avoided if it has wished.

Conclusion

.Lastly, the human body is a collection of vestigial and redundant structure and atavistic genes that serve no purpose but are occasionally expressed and can be the source of inconvenience such as the preauricular pit that can become infected but otherwise has no function, being an evolutionary remnant of the way our outer ear evolved from a gill arch in an ancestral fish, or some people's ability to waggle their ears the way ancestral simians could to give better detection of the direction sounds were coming from, or prick their ears when startled.

Likewise, the erector pili muscles that still try to fluff out our hair to keep us warm or make us look bigger to a threat, now serve no useful purpose. The additional cost in terms of resource to build the muscles, innervate them and the pathways to create and then metabolize the noradrenaline the nerve produce, has no detectable benefit.

Pseudogenes and 'junk' DNA are genetic remnants of past adaptations. Occasionally, these genes are repurposed through exaptation, demonstrating the evolutionary process of reusing existing resources.

Looked at objectively, beneath the superficial appearance of design, the human body, with its inefficiencies, vulnerabilities, and vestigial features, is best explained through the lens of evolution.

Far from reflecting intelligent design, our anatomy and physiology reveal a history of incremental changes shaped by natural selection and constrained by pre-existing structures. These imperfections underscore the reality of evolution as a tinkering process, producing functional but far-from-perfect outcomes. In this light, the human body stands as a powerful testament to our evolutionary heritage and tells a story far more impressive that the childish notion of it all being made by magic by a super-intelligent, yet invisible and undetectable designer.

The Body of Evidence

Appendix

The Body of Evidence

Appendix

Creationist claims rebutted

The following is a list of common creationist claims about the human body. As can be seen, they are based on a superficial understanding of the human body and ignorant incredulity, not on evidence or logical deductions.

The Human Eye

> Claim: The eye is irreducibly complex, perfectly suited for vision, and could not have evolved step by step.
>
> Rebuttal: The human eye exhibits significant flaws that are inconsistent with intelligent design:
>
>> Blind Spot: The optic nerve exits through the retina, creating a blind spot. This design is inefficient compared to cephalopods like octopuses, which have no such blind spot due to the absence of inverted retinal structures.
>>
>> Susceptibility to Disorders: Myopia, cataracts, and macular degeneration are common and highlight the vulnerabilities of the eye.
>>
>> Other Eyes: There are many other eyes in the natural world, some of which are superior to the human eye in terms of visual acuity and speed of image processing.
>>
>> Evolutionary Evidence: Simple light-sensitive cells in organisms like flatworms demonstrate plausible precursors to complex eyes. Gradual improvements over millions of years, such as the formation of lens structures in fish, provide a clear evolutionary pathway.
>
> Claim: Charles Darwin admitted that the eye is too complex to have evolved.

Rebuttal: This claim is based on a quote mine from *On The Origin Of Species* in which he mentions that perception of the created perfection eye as a common misconception, then followed it with a detailed explanation of how his theory could indeed explain it as the result of a gradual evolutionary process.

Blood Clotting Cascade

Claim: The complexity of the blood-clotting or coagulation cascade, involving numerous proteins and steps, could not have evolved through incremental changes.

Rebuttal:

Gradual Evolution: Simpler clotting systems exist in other organisms, such as lampreys, demonstrating intermediate stages. Each step likely conferred incremental survival advantages, such as better wound healing.

Flaws in the System: The cascade's complexity also makes it prone to errors like haemophilia or thrombosis, which would be unexpected in an intelligently designed system.

The unnecessarily complex nature of the coagulation cascade is characteristic of an evolved process. An intelligently designed process would be simpler and less prone to potentially fatal errors.

The Human Brain

Claim: The brain's complexity and capacity for abstract thought are evidence of purposeful design.

Rebuttal:

Appendix

Energy Demands: The brain consumes approximately 20% of the body's energy at rest, an enormous cost that leaves humans vulnerable to starvation and metabolic disorders.

Susceptibility to Disorders: Conditions like epilepsy, depression, and Alzheimer's disease reveal the brain's fragility and flaws, as does housing it in a skull without adequate protection against blows to the head, resulting in concussion or brain tissue injury.

The inability of the brain to repair injured tissues and the vulnerability of the brain to cope with elevated intracranial pressure are evidence of contingent, suboptimal evolution, not intelligent design.

Evolutionary Trade-offs: The expansion of the cerebral cortex for higher cognitive functions came at the expense of increased vulnerability to concussions and mental health issues.

Human Reproductive System

Claim: The intricacy of reproduction, from fertilization to childbirth, reflects intelligent design.

Rebuttal:

Difficult Childbirth: The narrow birth canal, a consequence of bipedalism, results in painful and often dangerous deliveries.

Male Reproductive Inefficiencies: The convoluted path of the vas deferens and the external placement of testes expose them to injury and temperature fluctuations, suboptimal designs inherited from evolutionary ancestors.

Evolutionary Constraints: These imperfections highlight the evolutionary tinkering of pre-existing structures rather than the implementation of a flawless design.

The Appendix

Claim: The appendix has immune functions, proving it is not a vestigial organ.

Rebuttal:

The fact of exaptation does not support intelligent design but instead illustrates how the process of evolution is constrained by having to make do with what is available.

Evolutionary Origins: The appendix evolved from a larger cecum used by herbivorous ancestors to digest cellulose. Its reduced size and function in humans make it a vestigial structure.

Susceptibility to Appendicitis: This life-threatening condition highlights the appendix's redundancy and potential danger, which are inconsistent with intelligent design.

The Immune System

Claim: The immune system's complexity is evidence of intelligent design.

Rebuttal:

Autoimmune Disorders: Conditions like lupus and rheumatoid arthritis illustrate the immune system's propensity to attack the body it is meant to protect.

Allergies: Hypersensitive immune responses to harmless substances, such as pollen, demonstrate inefficiency and maladaptation.

Appendix

Evolutionary Explanation: The immune system's complexity and flaws are the result of evolutionary pressures to combat pathogens, sometimes at the expense of self-tolerance.

Inefficiency: The immune system frequently fails to protect us against the pathogens that the same alleged designer also designed to overcome our immune system and make us sick.

Vestigial Structures

Claim: Structures like the coccyx and wisdom teeth have hidden functions, proving they are not vestigial.

Rebuttal:

Coccyx: While it provides minor support for pelvic muscles, the coccyx is primarily a remnant of a tail from primate ancestors.

Wisdom Teeth: These third molars often cause overcrowding, pain, and infections due to changes in jaw structure and diet over evolutionary time.

Vestigiality as Evidence: Vestigial structures are better understood as remnants of ancestral traits that have diminished in function, a hallmark of evolutionary processes.

Rationalisation: Assuming structures have a hidden function is not evidence that they have a function; it is evidence of self-delusion and the ability to rationalise unsupported beliefs.

The Heart and Circulatory System

Claim: The heart's efficiency and life-sustaining role indicate intelligent design.

Rebuttal:

Coronary Arteries: The susceptibility to blockages, leading to heart attacks, underscores a flawed design.

Arrhythmias: The heart's susceptibility to electrical instability resulting in life-threatening arrhythmias is evidence of evolved trade-offs balancing short-term efficiency with longer-term risks.

Human DNA

Claim: DNA is a complex code that must have been designed.

Rebuttal:

Non-Coding DNA: A significant portion of the genome consists of non-coding or "junk" DNA, much of which is vestigial or derived from viral insertions.

Redundant Genes: Many genes have been co-opted for new functions (exaptation), highlighting evolutionary ingenuity rather than intentional design.

Incredulity is not a scientific argument and there are several explanations for how the triplet code arose and why there could be no other.

Each of these claims, when examined through the lens of science, reveals the evolutionary history of the human body. The flaws, inefficiencies, and vestigial features present compelling evidence against the notion of intelligent design, instead supporting the process of natural selection and adaptation over time.

Index

References

1. **Rubicondior, Rosa.** *What Makes You So Special? From the Big Bang To You.* s.l. : CreateSpace Independent Publishing Platform (Amazon), 2017. ISBN-13 : 978-1546788294.

2. **Behe, Michael J.** *Darwin's Black Box: The Biochemical Challenge to Evolution.* 2nd. s.l. : Free Press, 2006.

3. **Darwin, Charles.** *On the Origin of Species by Means of Natural Selection or the Preservation of Favoured Races in the Struggle for Survival.* 1st. London : John Murray, 1850. Kindle Edition.

4. *Mysterious Indo-European homeland may have been in the steppes of Ukraine and Russia.* **Balter, Michael.** s.l. : The American Associaltion for the Advancement of Science, 13 February 2015, Science.

5. *Massive migration from the steppe is a source for Indo-European languages in Europe.* **Haak, Wolfgang, et al.** s.l. : Springer Nature Ltd, 2 March 2015, Vol. 522, pp. 207–211.

6. *Thermodynamics in landscape ecology: the importance of integrating measurement and modeling of landscape entropy.* **Cushman, Samuel A.** s.l. : Springer Nature Ltd, 6 November 2014, Landscape Ecology, Vol. 30, pp. 7-10.

7. **Behe, Michael J.** *Darwin Devolves: The New Science About DNA That Challenges Evolution.* s.l. : HarperOne, 2020. ISBN-13 : 978-0062842664.

8. **National Center for Science Education.** The Wedge Document. *NCSE.ngo.* [Online] National Center for Science Education, 14 October 2008. [Cited: 5 November 2024.] https://ncse.ngo/wedge-document.

9. **Discovery Institute.** The "Wedge Document" so what? *Discovery.org.* [Online] April 2019. [Cited: 5 November 2024.]

10. **Rubicondior, Rosa.** *The Unintelligent Designer: Refuting the Intelligent Design Hoax.* s.l. : CreateSpace Independent Publishing Platform, 2018. ISBN-13 : 978-1723144219.

11. **Various.** *Paleontology and Geology of Laetoli: Human Evolution in Context: Volume 2: Fossil Hominins and the Associated Fauna.* [ed.] Terry Harrison. s.l. : Springer, 2011. Vol. 2. ISBN: 978-9048199617.

12. *Craniofacial proportions in children with adenoid or adenotonsillar hypertrophy are related to disease duration and nasopharyngeal obstruction.* **Pawłowska-Seredyńska, Katarzyna, et al.** s.l. : Elsevier, May 2020, International Journal of Pediatric Otorhinolaryngology, Vol. 132.

13. **Rubicondior, Rosa.** Unintelligently Designed Teeth Cause Ray Discomfort. *Rosa Rubicondior.* [Online] 17 October 2013. [Cited: 14 December 2024.] https://rosarubicondior.blogspot.com/2013/10/unintelligently-designed-teeth-cause.html.

14. *Spatial mapping of polymicrobial communities reveals a precise biogeography associated with human dental caries.* **Kim, Dongyeop, et al.** [ed.] 12375-12386. 22, s.l. : The National Academy of Sciences, 18 May 2020, Proceedings of the National Academy of Sciences, Vol. 117.

15. *Caspase-11 mediated inflammasome activation in macrophages by systemic infection of A. actinomycetemcomitans exacerbates arthritis.* **Okano, Tokuju, et al.** 54, s.l. : Springer Nature Ltd., 15 August 2024, International Journal of Oral Science, Vol. 16.

Index

16. *Mucosal microbiome dysbiosis in gastric carcinogenesis.* **Coker, Olabisi Oluwabukola, et al.** 6, s.l. : BMJ, 1 August 2017, Gut, Vol. 67, pp. 1024-1032.

17. **Office for National Statistics.** Deaths due to and involving choking, by sex and age group, England and Wales: 2018 to 2022. *Office for National Statistics.* [Online] UK Office for National Statistics, 24 August 2023. [Cited: 14 December 2024.] Reference # 1438.
https://www.ons.gov.uk/peoplepopulationandcommunity/birthsdeathsandmarriages/deaths/adhocs/1438deathsduetoandinvolvingchokingbysexandagegroupenglandandwales2018to2022.

18. **Cancer Research UK.** What is Barrett's oesophagus? *Cancer Research UK.* [Online] Cancer Research UK. [Cited: 15 December 2024.] https://www.cancerresearchuk.org/about-cancer/other-conditions/barretts-oesophagus/about-barretts.

19. **Columbia Doctors.** History of Medicine: The Mysterious Appendix. *Columbia surgery.* [Online] Columbia University Erving Medical Centre. [Cited: 15 December 2024.] https://columbiasurgery.org/news/2015/06/04/history-medicine-mysterious-appendix.

20. **Rubicondior, Rosa.** *Unintelligently Designed Arms Races: How Nature Refutes Intelligent Design.* s.l. : Independently published, 2024. 979-8302686497.

21. *Denisovans and Homo sapiens on the Tibetan Plateau: dispersals and adaptations.* **Zhang, Peiqi, et al.** 3, s.l. : Elsevier Inc., March 2022, Trends in Ecology & Evolution, Vol. 37, pp. 257-267.

22. *Caspase-11 mediated inflammasome activation in macrophages by systemic infection of A. actinomycetemcomitans exacerbates arthritis.* **Okano, T., et al.** 54, s.l. : Springer Nature

Ltd., 15 August 2024, International Journal of Oral Science, Vol. 16.

Index

A

A back pocket in a vest as much sense as	113
Abdomen	22
Abdominal cavity	42, 49, 102, 111, 158
Abscesses	69
Acetabulum	43
Acetyl choline	79
Adenoid face	37
Adenoid tonsils	37
ADP	117, 118
Adrenal glands	56
Adrenaline	56, 79
Advanced economies	54
AF	See Atrial fibrillation
Africa	24, 50
African	
Elephant	15
savannah	44, 66, 77, 90, 111
Afrotheria	102
Allergic	
reaction	132
responses	131
Allergies	129, 131, 132, 133
Altaic Hypothesis	25
Alu elements	149
Alveoli	86, 90
Alzheimer's	167
Alzheimer's disease	119
Americas	24
Amino acid	18, 147
Amphibian ancestors	87
Amphibians	13
Ampulla of Vater	51
Amylase	149

Amylase gene...149
Anaemia 50, 72, 108
Anaesthetics..44
Anal canal...71, 72
Anaphylaxis..131, 132
Anatomy 40, 56, 64, 78, 93
Ancestral
    diet 68
    environment 60, 65, 82, 111, 128
    fish 161
    genes 146
    mammal 93
    mammals 111, 143
    remnants 145
    simians..161
    species..141, 146
    structures 144
    traits 145, 169
    vertebrates 62, 98
Angina 97
Ankle 44, 45
Anthropocentric arrogance 19
Anthropocentrism 7
Antiseptic surgery...44, 59
Anus 49, 71, 72
Anxiety...89, 122, 124
Aortic arch 152
Ape 23, 35, 36
Appendix 59, 156, 168, 173
Aquatic ancestors 39, 144
Argument from ignorant incredulity 8, 24, 75
Argument from perfection, The..8
Ark of the Covenant..71
Arms race 61, 105, 110, 114
Arrector pili muscle ....See Erector pili muscle
Arrhythmias 78, 79, 81, 84, 170
Arteries 76, 92, 95, 96, 98
Asia 24, 50
Asthma 88
Atavism..141, 145, 146
Atavistic
    fossil 59

Index

genes	160
structure	84
traits	141, 145, 147
Atelectasis	91
Atherosclerosis	53, 76, 95, 96, 98
Atlas	40, 41
ATP	117, 118
Atria	79, 80, 81, 82
Atrial fibrillation	81, 83
Atrioventricular node	80, 81, 82
Auricle	144

Auricular
- hillocks 144
- muscles 143

Australopithecine 22

Australopithecus
- afarensis 22
- sediba 16

Autoimmune
- conditions 53, 129, 131, 160
- diseases 61, 130, 131, 132, 133
- disorders 168

Autoimmunity 130, 132

Autonomic
- functions 120
- nervous system 56, 57

AV node .... See Atrioventricular node
Avulsion fracture 45
Axis 40

B

Bacteria 18, 36, 37, 38, 59, 61, 69, 86
Bad design 29, 67, 75
Barrett's Oesophagus 49
Basic life-support systems 120
Beauty and aesthetics 8
Behe, Professor Michael J. 9, 27, 28, 112
Benevolent designer 137
  inconsistent with the idea of a 133
Beta-blockers 82

Bible 71, 73, 80, 129, 207
Bifocal eyes 40
Big Bang..7, 208
Bilateral symmetry..20
Bile 51, 52
    duct 51
    salts...51
Biological resources 143
Biologically non-sensical 'devolution' excuse..................................71
Bipedal
    locomotion...72, 152
    posture 91, 92
Bipedalism16, 17, 22, 29, 35, 40, 44, 71, 73, 74, 91, 93, 94, 103, 104,
    111, 114, 117, 133, 136, 152, 153, 155, 159, 167
Bipolar disorder 124
Birds 10, 90, 142
Birth canal 17, 42, 103, 117, 156, 167
Blind spot...153
Blood pressure 57, 82, 97, 105, 106, 120
Bottlenose dolphin..15
Bowel
    movement 68, 71
    obstruction 55, 69
Bradycardia..81, 82
Brain15, 16, 17, 22, 35, 36, 37, 56, 65, 66, 76, 80, 89, 90, 105, 117,
    118, 119, 120, 121, 122, 124, 125, 126, 127, 128, 129, 136, 141, 150
    hemispheres of 126
    stem 36, 120, 121
Branchial clefts 144
Bronchi 88
Bronchioles...88
Bronchitis..86, 88
Bundle of His 80, 82
Bundle of Kent...80, 84

C

Caecum 59
Caesarean delivery 104
Calcaneus 45
Cambrian 20

Index

Carbohydrates 50, 53
Cardiac
    arrest 81, 82, 118, 160
    muscle 81, 159
    muscle cells 80
    output 120
Cardiomyopathy 82
Cardiovascular disease 113
Cardioversion 84
Carotid sinus 82
Cartilaginous fish 13
Cataracts 165
Cephalopods 10, 153, 165
Cerebella 15
Cerebrospinal fluid 121
Cervical curve 22, 40, 41
Cervix 103
Chemical imbalance 57
Childbirth 29, 42, 112, 113, 114, 133, 136
Childhood 17, 23, 37, 61
Chimpanzees 22, 66, 119, 143
Choking 39, 40, 49, 85, 86, 158, 173
Cholesterol 51, 52
Chordates 20, 120
Christian 27, 112, 129
    fundamentalism 112
    theology 27
Chromosomal abnormalities 135
Chronic Obstructive Pulmonary Disease 88
Chronic traumatic encephalopathy 122
Circular reasoning 17, 21
Circulatory
    failure 96
    system 75, 76, 85, 86, 92, 118
Circumflex artery 95
Clonic phase 126
Clotting factors 13, 14, 75, 78
Coagulation cascade 11, 12, 13, 75, 166
Coccydynia 142
Coccyx 141, 142, 156, 169
Co-evolution 17
Co-evolved symbiotic relationship 61

Cognition 117
Cognitive
    ability 122
    capabilities 119, 120
    decline 122
    skills 23
Cognitive ability 160
Colon 60, 63, 68, 69, 70
    cancer 60
Comfort, Ray 38
Common cold 37
Communication 16, 17, 23, 40
Complex
    cognitive functions 118
    speech 49, 85
    vocalizations 40
Compromises 23, 28, 29, 35, 52, 64, 66
Concussion 121, 122, 125, 160, 167
Coning 120, 121, 125
Consciousness 90, 118, 128, 160
Constipation 50, 64, 71, 72
Coronary
    arteries 170
    artery 81, 95, 96, 97, 98
    circulation 95, 97, 98, 159
    diseases 98
    system 98, 99
Cranium 16, 36, 120
Creationism 8, 20, 27, 208
Creationist7, 8, 10, 11, 14, 16, 26, 27, 30, 71, 73, 75, 80, 97, 134, 147, 165, 210
    belief 134
Creativity 124, 125, 128
Creator god 80
Crebral hemispheres 120
Crohn's disease 60
Cruciate ligaments 44
C-section 104
Cultural
    difference 123
    evolution 17, 73
Cushman, Samuel A 27

Index

Cytokines 56

D

Darwin, Charles 10, 165
Darwinian 9, 14, 28, 39
   evolution 9
Death 50, 52
Deep vein thrombosis 76, 77, 94
Defibrillation 84
Defibrillator 81
Dehydration 60, 110
Déjà vu, ...126
Deleterious mutation 27
Dementia 122
Denisovan 16
Dens 40
Dental
   carries 38
   plaque 38
Depression 122, 124, 167
Design
   errors 59
   flaw 38, 49, 65, 68, 88, 93, 95, 96
   poorly optimized 137
Designer god 7, 8, 26, 71, 75
Devolution 27, 28, 71, 112
Diabetes 52, 53, 54, 55, 57, 64
   type I 130, 160
Diaphragm 49, 50, 87
Diarrhoea 60, 64, 67
Diastole 96, 97
Diastolic pressure 96, 97
Diet 36, 49, 52, 54, 59, 60, 61, 65, 66, 68, 69, 72, 73, 96
Digestion 50, 57, 59, 62
Digestive
   enzymes 50, 51, 59
   juice 50
   system 51, 55, 56, 57, 58, 64, 67, 68, 75, 85, 91
   tract 49, 62, 68, 70, 72, 75, 142
Discovery Institute 27, 28

Disease 53, 55, 63, 64, 70, 85, 136, 172
Diverticula 68, 69, 70
Diverticulitis 68, 69, 70
Diverticulosis 69
DNA 14, 18, 28, 113, 134, 145, 147, 149, 150, 161, 170, 171
   non-functional 150
   replication 28
   replication process 113
Dogma 27
Drainage tubes 36
Duodenum 51, 52
Dust mites 131
DVT ....See Deep vein thrombosis

E

Ear 37, 144, 145
East Africa 21
East African savannah 68
Ectopic pregnancy 103, 114
Electrolyte imbalances 58, 110
Elephant 15, 102
Elephants 158
Embryogenesis 144
Empathy 16
Emphysema 88
End arteries 95, 96
Endogenous Retrovirus 148
Endometrium 104, 107, 109, 135
Endothelial cells 77
Enhancers 150
Environment 16, 17, 21, 23, 37, 54, 61, 63, 64, 66, 67, 70, 77, 85, 86, 87
Environmental
   factors 60, 65
Enzyme 11, 13, 18, 50, 147, 150, 151
Epididymis 43
Epiglottis 39, 158
Epilepsy 125, 126, 127, 128, 129, 167
   temporal lobe 129
Epileptic
   seizures 126, 159

Index

Epileptic seizures 126
Epithelial cells..50
Epitome of stupidity 105
Erector pili muscle 143, 161
Error-prone
    complexity 75
    solution..67
ERVs ....See Endogenous Retroviruses
Eukaryote 18
Europe 24, 50
Eustachian tube..37
Evolution7, 9, 10, 13, 14, 16, 17, 19, 21, 22, 27, 28, 35, 38, 39, 40, 42, 44, 49, 52, 54, 55, 57, 58, 60, 62, 63, 64, 66, 67, 68, 69, 70, 72, 73, 74, 75, 77, 78, 83, 84, 85, 88, 91, 92, 93, 94, 95, 98, 99, 102, 103, 104, 106, 108, 109, 111, 112, 113, 114, 118, 120, 121, 125, 129, 130, 131, 132, 133, 134, 136, 137, 139, 142, 144, 145, 147, 148, 149, 150, 151, 152, 154, 156, 158, 166, 171, 172, 173
    compelling evidence for 113, 123, 133, 141
    compromises inherent in 134
    constraints and compromises of 127
    constraints of..103
    evidence for 125
    observable evidence of 147
    patchwork nature of 120
    process of 117
    strong evidence for 129
    tinkering nature of 145
    trade-offs inherent in 120
Evolutionary
    adaptation 92, 93, 104, 108, 110
        limitations of..58, 64, 73
    advantages 94, 124
    ancestors 134, 167
    arms race 105
    baggage 142
    benefits 56
    cause 96
    changes 103
        incremental..66
    compromise 42, 67, 82, 114, 123
    compromises...156, 157, 158
    consequences 145

constraints 64, 92, 112, 121, 159, 168
contingency 107
evidence 165
explanation 8, 69, 97, 102, 113, 169
history 21, 40, 49, 53, 66, 70, 75, 80, 85, 113, 144, 146, 147, 154, 156, 170
inefficiencies 70
ingenuity 170
leftovers 143
mismatch 124, 142
need 136
optimization 93, 97
origin 152, 153
origins 84, 95, 101, 144, 159, 168
past 141, 145
pathway 165
pressure 29, 44, 51, 73, 92, 93, 94, 98, 102, 107, 110, 134
pressures 169
priority 96
process 7, 11, 14, 16, 30, 52, 57, 58, 60, 61, 68, 75, 77, 85, 87, 91, 98, 112, 125, 127, 137, 146, 151, 155, 157, 159, 160, 161, 166
  consistent with an 133
Process 129, 169
reason 103
remnant 161
solutions 151
strategy 135
time 156, 169
tinkering 101, 168
trade-off 167
trade-off 55, 60, 65, 74, 95, 109, 114, 118, 119, 121, 128, 132, 133, 151
transition 144
trial and error 135
Evolved 9, 10, 12, 13, 14, 16, 23, 24, 27, 28, 35, 39, 40, 41, 42, 43, 54, 55, 57, 58, 61, 62, 66, 73, 75, 77, 82, 83, 85, 86, 87, 90, 93, 94, 95, 96, 97, 98, 101, 102, 103, 104, 105, 106, 107, 108, 109, 110, 112, 113, 120, 122, 123, 124, 125, 127, 129, 130, 131,141, 144, 146, 150, 152
  mechanism 56
  organs 80
  process 13, 166

Index

Exaptation 9, 39, 59, 127, 151, 161, 168, 170
Exapted 39, 147, 149, 150
External ear 144
Eye 10, 18, 36, 142, 152, 153, 165, 166
Eyeball 142

F

Facial architecture 38
Factor II 12
Factor IIa 12
Factor IX 12, 13
Factor IXa 12
Factor V 12
Factor Va 12
Factor VII 12
Factor VIII 12, 13
Factor VIIIa 12
Factor X 12, 13
Factor Xa 12
Factor XI 12
Factor XIa 12
Factor XII 12
Factor XIIa 12
Factor XIIIa 12
Fallen arches 45
Fallopian tubes 103
False dichotomy fallacy 9
Fatal errors 166
Fearful infancy of our species 112
Femur 44
Fibrinogen 12, 13
Fight or flight
    diarrhoea 66
    response 56
Fine tuning and purpose 8
Fistulas 69
Flawed system 67, 94
Focal seizures 126
Foetal heamoglobin 149
Foetus 105, 106, 110, 111, 112, 114, 117, 148, 149

Foramen magnum 22, 36, 120
Four-chambered heart..83
FOXP2..150
Functional but flawed 67, 74
Functional genes 148
Fundamentalist...207

G

Gall
   bladder 51, 52
   stones 51, 52
Gamete production 133
Gametes...136
Gastric ulcers 50, 51
Gazelle 23
Gene 13, 18, 24, 27, 54, 75, 87, 96, 113, 128, 130, 134, 136, 146, 147,
   148, 149, 150, 151, 161, 170
   duplication 13, 75, 149
   pool 27, 54, 136
   remnants of 148
Generalized seizures..126
Genetic 13, 21, 27, 28, 52, 53, 60, 71, 76, 83, 91, 124, 128, 130, 132,
   135, 136
   disorders 28
   diversity...136
   drift...145
   elements 151
   entropy...27, 71
   information 147
   material 148
   mutation 76, 83
   mutations 129
   pathways...146
   predisposition 53, 60
   remnants 161
Genetics 96
Genome...19, 134, 149, 170
Gill arch 39, 144, 161
Giraffe..152
Globin gene 149

Index

Globins 149
Glucagon 53, 54
Glucose 117, 118, 120
God of the gaps 9
Golden eagle 10
Good design 29, 72
Goosebumps 143
Gorillas 119, 143
Grand mal 126
Grandmother hypothesis 113
Great apes 151
Group
   cohesion 23, 108
   norms 23
GULO 150
Gut-brain axis 56, 65, 66

H

Haem 51
Haemochorial 104
Haemophilia 76, 166
Haemorrhoids 70, 71, 72, 73
Hair follicles 143
Half-wit designer 91
Hallucinations 126, 129
HARE5 150
hCG .... See Human chorionic gonadotrophin
Heamoglobin 149
Heart 49, 53, 56, 76, 78, 79, 80, 81, 82, 83, 84, 87, 92, 94, 95, 96, 97, 98
   attack 95, 97
   block 82
   failure 97
   muscle 95
   rate 79
Heath-Robinson 62, 102
   contraption 62
Helicobacter pylori 50, 51
Helminths 131
Herniated discs 152
HG .... See Hyperemesis gravidarum

Hiatus hernia 49, 50
Hip joint 43
Hippocampi 15
HLA gene family 130
Homeobox genes 18
Homeostasis 66, 75
Hominin 16, 22, 61, 93, 96
Homo sapiens 16, 36, 57
Hormonal imbalances 113
Hox genes .... *See* Homeobox genes
Human
    biology 60
        messy nature of 111
    birth canal 136
    body 7, 8, 10, 14, 20, 28, 30, 44, 55, 56, 58, 64, 66, 70, 71, 73, 84,
        91, 94, 98, 141, 155, 159, 160, 161, 165, 170
    brain 15, 17, 104, 117, 118, 120, 121, 123, 159
    childbirth 137
    chorionic gonadotrophin 110, 111
    communication 150
    culture 7
    evolution 16, 109
    genome 147, 148, 151
    menstrual cycle 135
    reproduction 136
    reproductive strategy 133
    sexuality 108
    spine 152
Humans 7, 10, 14, 15, 16, 17, 19, 20, 24, 39, 40, 42, 49, 50, 52, 58, 63,
    64, 66, 69, 72, 74, 76, 85, 86, 87, 96, 104, 108, 111, 118, 119, 133,
    134, 137, 141, 142, 143, 145, 146, 149, 150, 151, 152, 208
Hunter-gatherer 22, 54, 72, 90, 93
    communities 124
Hydrochloric acid 50
Hyperallergic reaction 37
Hyperemesis gravidarum 109, 110, 111, 112
Hyperglycaemia 53
Hyperkalaemia 82
Hypermania phase 124
Hypertrichosis 146
Hyperventilation 88
Hypoglycaemia 119

Index

Hypothermia 82
Hypothyroidism 82

I

IBD .... See Inflammatory bowel diseases
IBS .... See Iritable bowel syndrome
Ignorant incredulity 165
Immune
   function 149
   response
      gene 149
      responses 148, 168
         poorly regulated 61
   system 14, 27, 52, 53, 59, 61, 62, 63, 64, 88, 129, 130, 131, 132, 133, 160, 168, 169
      failure of 132
      tolerance 62
Immunity 148
Implantation 103, 104, 106
Increased libido 113
Incredulity 170
Indo-European Languages 24
Infarction 81
Infection 36, 37, 38, 50, 51, 55, 56, 57, 59, 60, 61, 69, 86, 91
Inflammation 56, 57, 58, 60, 64, 65, 69, 70, 88, 96
Inflammatory
   bowel diseases 60
   disorders 61
Information 18, 20, 23, 124, 147
Inguinal
   canal 43
   hernia 43
Insulin 52, 53, 54
Intellectual gymnastics 132
Intellectually bankrupt 14
Intelligent creator 91, 95
Intelligent design 9, 19, 23, 27, 28, 39, 52, 57, 58, 60, 63, 64, 68, 69, 70, 72, 74, 75, 77, 78, 84, 85, 87, 92, 94, 98, 99, 101, 106, 109, 112, 113, 117, 121, 125, 127, 129, 133, 136, 137, 141, 142, 151, 152, 153, 155, 160, 161, 167, 168, 169, 170

advocates 63, 65, 94, 114
challenging the notion of 151
compelling argument against 55, 67, 78, 84, 92, 133
compelling evidence against 113, 125
incompatible with notions of 136
inconsistent with 126, 165, 168
inconsistent with the notion of 105, 108, 135
process
    utterly inconsistent with an 133
refutes any notion of 132
refuting the notion of 145
the antithesis of 127, 143
vulnerabilities that refute any notion of 109
Intelligent designer14, 28, 38, 40, 51, 52, 53, 54, 59, 60, 61, 62, 63, 64, 67, 68, 70, 72, 78, 79, 83, 84, 86, 88, 90, 91, 93, 94, 95, 96, 97, 98, 99, 101, 103, 104, 105, 109, 111, 113, 114, 120, 121, 132, 134, 135, 136, 141, 145, 153, 154, 158, 160
    difficult to reconcile with the idea of an 114
    incompatible with the idea of... 147
    inconsistent with the notion of 105
    particularly incompatible with the notion of an 131
    putative 63
Intelligent involvement
    lack of... 101
Intelligent solution.. 103
Intelligently designed18, 24, 28, 62, 71, 73, 76, 78, 80, 91, 102, 105, 106, 108, 119, 122, 124, 128, 132, 154, 157, 166
    if the system were 143
Intelligently designed system 76, 78
Intercostal muscles... 49
Intercourse 107, 108, 109, 113
Interferon 149
Interventionist god 123
Intervertebral discs 41
Intraluminal pressure 68
Introspection 124, 125
Invertebrates 12
Iron-deficiency anaemia 107
Irreducible complexity.. 8, 9
Irreducibly complex 10, 14, 75, 165
Irritable bowel syndrome 64, 65, 66, 67
Ischemia 95

Index

Islets of Langerhans 52
Israelites 71

J

Jaw 36, 38
Jawless fish 13
Jellyfish 77
Jerry-built systems 85
Junk gene 148

K

Ketones 118
Kudu 23

L

LAD .... See Left anterior descending artery
Laetoli footprints 35
Lampreys 13, 166
Large intestine 59, 64, 68
Larynx 39, 40, 152, 158
Late Bronze Age 112
Lateral malleolus 45
Laws of Thermodynamics 147
Layer of complexity 13, 28, 127
LCA .... See Left coronary artery
Left anterior descending artery 95, 97
Left atrium 79
Left coronary artery 95
Left ventricle 79
Leopard 23, 24, 90
Life-threatening 38, 43, 55, 58, 60, 69, 70, 76, 88, 90, 94, 95, 98, 103, 105, 110
Limbic system 15
Lion 23
Locally-popular god 7, 8, 9, 17, 123

Lumbar
  curve 22
  vertebrae 41
Lupus 130, 160, 168
Lymphatic tissues 59

M

Macular degeneration 165
Malabsorption 60, 67
Maladaptation 168
Maladaptive 57, 65, 66, 110, 112, 124, 132
  response 57
Malevolent 38, 73
Mammalian
  ancestors 144, 146
  ancestry 146
  family tree 151
  heart 78, 82
Mammals 13, 39, 40, 41, 42, 43, 49, 91, 92
  ancestors of 142
  terrestrial 15
Manatees 102, 158
Mandible 38
Marine vertebrates 39
Mastoid abscess 37
Maxilla 38
Maximal simplicity 133
Medial malleolus 45
Meningitis 37
Menisci 44
Menopause 112, 113
Menstruation 107, 108, 109, 137
  evolution of 108
Mental gymnastics 68
Mental health 121, 123, 124, 125, 159, 160, 167
  disorders 122, 123
Mentruate 109
MER41 149
Metabolic costs 119
Metabolic pathway 151

Index

Microbes	61, 62
Microbiome	59, 61, 62, 63, 65
Microbiome-dependent digestion	66
Microorganisms	59
microRNA	148
Middle ear	37
bones	152
Mindless evolution	59
Minimal waste	133
Miscarriage	135, 137
Modern environments	96, 130, 131
Monogamy	108
Morning sickness	109, 110
Motor control	87, 126, 128
Multicellular organisms	18
Multiple sclerosis	130
Mutation	27, 28, 83
Myocardial infarction	76, 95
Myocarditis	82
Myocardium	95
Myopia	165

N

Narrative	23
Nasal cavity	36
Nasopharynx	37
Natural	
pacemaker	79, 81
process	30, 40, 53, 63
selection	14, 29, 91, 95, 98, 120, 125, 128, 129, 132, 133, 135, 136, 137, 145, 147
selectors	23
Natural selection	155, 161, 170
Nausea	109, 110, 111, 112
Neanderthals	16, 35
Needless complexity	67
Neocortex	15, 150
Neonatal respiratory distress syndrome	90
Neural	
activity	128

circuitry	124
mechanisms	124
Neurodegeneration	122
Neurodegenerative diseases	119
Neurological disorder	125
Neuronal	
connections	122
ion gradients	117
Neurotransmitters	124
dysregulation of	127
New Testament	129
Nictitating membrane	10, 142
Non sequitur	24
Noradrenaline	56, 143, 161
Norepinephrine	See Noradrenaline

O

Obesity	53, 54, 71, 93
Obstructive jaundice	51
Occam's Razor	17, 19
Oesophageal cancer	49
Oesophagus	39, 49, 85
Old Testament prophets	123
Olfactory lobes	15
Omnibenevolent	
creator	105
designer	133
god	141
Omnibenevolent designer	159
Omniscient designer	
compelling argument against the notion of an	70
difficult to reconcile with the idea of a	121
Oocytes	135
Opportunist bacteria	37
Opposable thumb	16
Optic nerve	153, 165
Origin myth	23, 112
Osteoporosis	44, 113
Otitis media	37
Outer ear	144

Index

Ovaries 102
Overactive immune response...60
Overly complex...66, 77
Ovulation 135

P

Pair-bonding 113
Palpitations..81
Pancreas 51, 52
Pancreatic duct..51
Pancreatitis..51
Paralytic ileus 55, 56, 57, 58, 64
Parasite 21, 27, 61, 62, 71, 105, 130, 131
Parasitic...105, 112, 114, 148, 149
   arms race 112
   organism 105
Partial pressure 86
Past environment 69
Pathogen 50, 61, 66, 86, 107, 108, 109, 110, 130, 131, 132
Pathogens 158, 160, 169
Pelvis 22, 42, 103, 104, 114
Peregrine falcon 10, 142
Perfect designer
   challenge to the concept of a 128
Perinatal mortality..104
Peristalsis 55, 56, 57, 58
Peritoneum 59
Peritonitis...55, 56, 69
Petit mal 126
Pharyngeal arches 144, 152
Pharynx 39, 40, 49, 85, 158
Philistines 71, 73
Phospholipids 51
Phrenic verve 87
Physical evidence..141
Physiology 14, 53, 56, 70, 92, 93
Piles..70, 71
Placenta 103, 104, 105, 110
Placenta previa...103, 104, 105, 106, 114
Plantar fasciitis 45

195

Plaque 76, 96, 97
Platelets 11, 75, 77
Pneumonia 86
Pollen 131
Polygamous sex 108
Polygamy 108
Polythelia 146
Post-ictal phase 126
Preauricular
    pits 144, 145, 161
    sinus .... See Preauricular pits
Predation 136
Pre-eclampsia 105, 106, 114
Pregnancies 103, 104, 135
Pregnancy50, 71, 104, 105, 106, 107, 109, 110, 111, 112, 113, 117, 133, 134, 136, 137
Pre-hominin ancestry 61
Pre-mammalian ancestor 55
Primate ancestors 141, 156, 169
Procreation 108, 113
Prokaryotes 18
Prolapsed
    haemorrhoids 71
    uterus 42
Prolific waste 133
Prostaglandins 56
Prostate 102, 113
    cancer 102, 103
    gland 102
Prostatic hyperplasia 102
Protein C 12, 13
Proteins 11, 12, 13, 18, 50, 75
Prothrombin 12
Pseudogene 148, 150
Pseudogenes 148, 161
Psychosis 124
PTEN 148
PTENP1 148
Pulmonary
    circulation 79, 94
    embolism 76, 77, 94
    oedema 79, 86

Index

system	85, 86, 88, 91
Purkinje fibres	80, 82
Python	24, 85

Q

Quadrupedal
- ancestry 152
- animals 41
- structure 152

Quadrupeds 43, 72, 94, 104, 152

R

Radial ligaments 45
RCA .... See Right coronary artery
Rectum 60, 71, 72
Recurrent laryngeal nerve 152
Redundant
- coding 147
- gene 147

Redundant genes 170
Regeneration 122
Religious
- experience 129
- fundamentalist 9, 108, 113
- opinion 134

Reproductive
- advantage 133
- biology 108
- cancers 113, 114
- fitness 108
- physiology 134
- process 135
- strategy 133, 136, 137
- success 63, 70, 73, 83, 98, 112, 113, 118, 128, 134, 136
- system 101, 103, 104, 106, 112, 114, 133, 137
    - female 101
    - human 106, 113, 114
    - male 101

197

tract
    female 102, 107, 109, 135
Reptiles 13, 120, 142
Resource scarcity 136
Respiration 49, 87, 89
Respiratory systems 87, 158
Respiratory tidal volume 120
Rheumatoid arthritis 38, 130, 160, 168
Right atrium 79, 80
Right coronary artery 95
Right ventricle 79
RNA 18
Running 14, 42, 44, 45, 89, 90

S

SA node .... See Sinuatrial node
Saul of Tarsus 129
Savannah 22, 23, 35, 54
Schizophrenia 122, 123, 124
Schizophrenic
    voices 123
Sciatic nerves 41
Scoliosis 152
Scrotum 43, 102
Selection 17, 22, 145
Selection pressure 17, 22, 159, 160
Semi-lunar cartilages 44
Septum 79, 80
Sexual
    activity 107, 108
    intercourse 108
    receptivity 109
Sexuality 109
Sexually active 113
Sharks 13
Sick Sinus Syndrome 82
Sigmoid colon 68
Simian ancestor 22, 151
Simians 36
Single-celled organisms 86

Index

Sino-Tibetan Languages 25
Sinuatrial node 82
Sinus infections 37
Sinuses 36, 37
Skull 22, 35, 36, 41
Slipped disc 41
Small intestine 43
Smoking 96
Social
    bonding 107, 108
    cohesion 107, 113
    ethics 23
    structure 15, 16, 23
South Africa 16
Sparrowhawk 10
Special creation 16
Sperm 15, 16, 43, 101, 102, 134, 135, 136
Sperm whale 15
Spinal column 22, 36, 41
Spinal cord 36, 41
Spine 22, 41, 42, 43, 141, 152
Steppe Hypothesis 24
Stillbirths 135
Stomach 38, 49, 50, 51
    cancer 38
Story-telling ape 23
Straw man 28
Streptococcus mutans 38
Stress 42, 43, 50, 55, 56, 57, 60, 65, 66, 67, 82, 83, 88
    incontinence 42
Stress-induced gut responses 66
Stroke 76, 119, 125
Structural
    abnormalities 83
    defects 72
    integrity 62
Suboptimal 13, 29, 30, 39, 42, 44, 54, 57, 58, 68, 70, 72, 75, 79, 85, 92, 102, 114, 130, 151, 154
    compromise 39, 42, 44, 68, 70, 75
    design 39, 40, 72, 130, 154, 158, 167
    evolution 167
    outcomes 114

    process 13, 58
    solution 29, 30, 57, 85
    system 92
    workaround 102
Suden cardiac death 81
Superficial resemblance of design 28
Symbiotic
    association 112
    benefits 50
Sympathetic nerve 143
Synaptic activity 117
Syncytiotrophoblast 148
Systemic
    circulation 79
    inflammation 60
Systole 97

T

Tachycardia 81
Talus 45
Teeth 29, 35, 36, 38, 49
Teleological thinking 24
Temporal lobe 15, 129
Terrestrial tetrapods 39
TEs .... See Transposable elements
Testes 43, 101, 102, 158, 167
Testicular torsion 101, 102
TF-VIIa complex 12
Theory of Evolution 10, 17
Third eyelid 142
Thoracic
    cavity 49
    vertebrae 41
Thrombin 12, 13
Thrombophlebitis 92
Thrombosis 13, 72, 75, 76, 77, 78, 94, 158, 166
Tibia 44, 45
Tissue factor 12
Tonic phase 126
Tower of Babel 26

Index

Toxic megacolon	60
Toxins	110
Trachea	39, 85
Transcriptase	18
Transposable elements	149, 150
Traumatic brain injury	121
Trepanning	120
Trigeminal nerve	37
Triplet code	170
Trochanter	44
Tuberculosis	86
Tumour suppressor gene	148
Type III Secretory System (T3SS)	9

U

UC	*See* Ulverative colitis
Ulcerative colitis	60, 61, 62, 63, 64, 65
Ulcers	50, 64, 92, 94
Unconsciousness	119
Unintelligent design	51
Unintelligently designed	75
Universe	7, 15, 141
Unnecessarily complex	166
Unnecessary	
complexities	101
complexity	133
waste	67
Upright posture 22, 35, 40, 41, 42, 49, 70, 91, 93, 103, 114, 117, 133, 152	
Uralic Languages	25
Urethra	101, 102
Urinary	
obstruction	102
retention	103
tract infections	102
Uterine health	108
Uterus	103, 104, 107, 111, 135
UTI	*See* Urinary tract infections

V

Vagus nerve 56, 79, 82
Varicose veins 92, 93, 94, 95, 159
Vas deferens 101, 114, 153, 167
Venous
    insufficiency 92, 93
    system 94
Ventricles...79, 80, 81, 82
Ventricular
    fibrillation 81, 84
    tachycardia...81
Vermiform appendix 59
Vertebral column 40
Vertebrate
    ancestors 114
    body 75
    evolution 101
Vertebrates...10, 13, 43, 77, 82, 83, 85, 87
    cold-blooded 102
Vestigial
    feature 144, 145, 161, 170
    organs 27, 141, 147
    remnants 144
    structure 142, 145, 168, 169
Victim blaming 112
Viral gene 148
Virus infection 37
Visceral hypersensitivity 65
Visual acuity 11, 165
Vitamin C 150, 151
Vomiting 109, 110, 111, 112
von Willebrand factor...11
Vulnerabilities55, 57, 60, 63, 69, 71, 76, 84, 85, 95, 98, 99, 101, 109, 114, 119, 120

W

Walking..21, 23, 42, 44, 45, 93, 94
Wedge Strategy...28

Index

Whales 15, 16, 85
Whiplash injury 41
Wisdom teeth 38, 142, 156, 169
   impacted 142
Wolfe-Parkinson-White syndrome 79, 80, 83, 84
WPW syndrome .... See Wolfe-Parkinson-White syndrome

Y

Yamnaya 24
Young Earth creationist 38

Z

Zygote 20, 103

The Body of Evidence

Other Books by Rosa Rubicondior

(Prices correct at time of publication. Check online for current details)

The Light of Reason Series:

The Light of Reason: And Other Atheist Writings.

Irreverent essays, thought-provoking articles and humorous items on atheism, religion, science, evolution, creationism and related issues.

(Hardcover|) ISBN-13: 979-8512173916	£13.75 (US $18.75)
(Paperback) ISBN-10: 1516906888, ISBN-13: 978-1516906888	£9.20 (US $12.75)
(Kindle) ASIN: B014N0IPVI	£5.50 (US $7.50)

The Light of Reason: Volume II – Atheism, Science and Evolution.

Thought-provoking essays on the conflict between fundamentalist religion and science, and exposing the anti-science, extremist political agenda of the modern creationist industry.

(Hardcover) ISBN-13: 979-8512191040	£13.75 (US $18.75)
(Paperback) ISBN-10: 1517105188, ISBN-13: 978-1517105181	£9.45 (US $11.75)
(Kindle) ASIN: B014N0IR16	£3.99 (US $5.99)

The Light of Reason: Volume III – Apologetics, Fallacies, and Other Frauds.

Thought-provoking essays and articles on religion and atheism, dealing with religious apologetics, fallacies, miracles and other frauds

(Hardcover) SBN-13: 979-8512526002	£13.90 (US $17.25)
(Paperback) ISBN-10: 151710761X, ISBN-13: 978-1517107611	£7.75 (US $10.75)
(Kindle) ASIN: B014N0IRE8	£3.50 (US $5.50)

The Light of Reason: Volume IV - The Silly Bible.

Exposing the absurdities, contradictions, and historical inaccuracies in the Bible and advancing the case for atheism and against religion. This volume, the fourth in the Light of Reason series, deals with contradictions and absurdities in the Bible.

(Hardcover) ISBN-13: 979-8512539392	£13.75 (US $18.75)
(Paperback) ISBN-10: 1517108209, ISBN-13: 978-1517108205	£8.22 (US $10.20)
(Kindle) ASIN: B014N0IR8E	£3.99 (US $4.99)

The Light of Reason: And Other Atheist Writing. (all 4 volumes in one book)

Based on the Rosa Rubicondior science and Atheism blog, this is a collection of Atheist and science articles, some short, others lengthier, exploring the interface between religion and science and which have been published over some four years.

(Kindle only) ASIN: B013DYOK32   £6.34 (US $9.95)
(Paperback) ISBN-13: 978-1521146330   £24.00 (US $30.50)

Other books on science, Atheism and theology

An Unprejudiced Mind: Atheism, Science & Reason.

Essays on science and theology from a scientific atheist perspective, exploring particularly evolution versus creationism.

(Hardcover) ISBN-13: 979-8512554685   £13.35 (US $18.75)
(Paperback) ISBN-10: 1522925805, ISBN-13: 978-1522925804   £8.75 (US $11.75)
(Kindle) ASIN: B019UGXPM4   £3.99 (US $5.95)

Ten Reasons To Lose Faith: And Why You Are Better Off Without It.

Why faith is not only a fallacy and useless as a route to the truth but is actually harmful to society and to the individual. It systematically dismantles the standard religious apologetics and shows them to be bogus and deliberately constructed to mislead.

(Hardcover) ISBN-13: 979-8509108433   £18.90 (US $24.00)
(Paperback). ISBN-13:978-1530431953, ISBN–10: 1530431956   £12.60 (US $16.00)
(Kindle) ASIN: B01DGVO3JS   £6.90 (US $9.50)

What Makes You So Special? : From the Big Bang to You.

How did you come to be here, now? This book takes you from the Big Bang to the evolution of modern humans and the history of human cultures

(Hardcover) ISBN-13: 979-8509108433   £14.45 (US $18.40)
(Paperback) ISBN-13: 978-1546788294, ISBN-10: 1546788298   £10.00 (US $12.90)
(Kindle).ASIN: B071FTKXLZ $5.69 (US $7.99)

Refuting Creationism: Why Creationism Fails In Both Its Science Its Theology

Creationism is not science and has no scientific validity. It' claims are untestable and free from supporting evidence and it makes no useful predictions. Arguments for it are invariably presupposition arguments from ignorant incredulity, god of the gaps and false dichotomy fallacies.

(Hardcover) ISBN: 979-8345634912   £14.13 (US $18.00)
(Paperback) ISBN: 979-8345104989   £9.81 (US $12.50)
(Kindle) ASIN B0DM2N14VW   £5.00 (US $6.50)

Other Books By Rosa Rubicondior

The Failure of Creationism: The Theory That Never Was

Why the pseudoscience of creationism had failed to make any significant inroads into mainstream science

(Hardcover) ISBN: 979-8301041938 £16.00 (US $20.00)
(Paperback) ISBN: 979-8300579050 £10.00 (US $12.50)
(Kindle) ASIN: B0DNWPSKR1 £6.00 (US £7.50)

Unintelligently Designed Arms Races: How Nature Refutes Intelligent Design

The evolutionary arms races between species and even withing species that pervade nature are completely incompatible with the notion of intelligent design.

(Hardcover) ISBN: 979-8302694881 £14.20 (US $18.00)
(Paperback) ISBN: 979-8302686497 £8.30 (US $10.50)
(Kindle) ASIN: B0DPRH2ZVZ £5.35 (US $6.75)

A History of Ireland: How Religion Poisoned Everything.

From the earliest beginnings to the Northern Ireland 'Troubles' and beyond. Religion has had a major role in spreading divisions and providing excuses for subjugation and repression. Only rarely has religion played a constructive role in the development of Irish culture and political life.

(Hardcover) ISBN-13: 979-8507235032 £15.75 (US $20.00)
(Paperback): ISBN-13: 978-1724988492 £9.50 (US $12.10)
(Kindle) ASIN: B07HHHRB34 £6.10 (US $7.25)

The Internet Handbooks series

The Internet Creationists' Handbook: A Joke for the Rest of Us.

A humorous look at creationist apologetics on the Internet, showing the fallacies and dishonest tactics creationists are using to try to recruit scientifically illiterate people into their political cult.

(Paperback),ISBN-13: 978-1721605149, ISBN-10: 1721605149   £5.78 (US $7.75)
(Kindle) ASIN: B07DZF75KD   £3.75 (US $5.00)

The Christian Apologists' Handbook: A Joke for the Rest of Us.

A humorous look at Christian apologetics on the Internet, showing the fallacies and dishonest tactics Christian fundamentalists are using to try to recruit scientifically and theologically illiterate people to their cults, often with political motives.

(Paperback) ISBN-13: 978-1721724727, ISBN–10: 1721724729   £6.25 (US $7.75)
(Kindle) ASIN: B07DYDVMW4   £3.75 (US $5.00)

The Muslim Apologists' Handbook: A Joke for the Rest of Us.

A humorous look at Muslim apologetics on the Internet, showing the fallacies and dishonest tactics Muslim fundamentalists are using to try to recruit scientifically and theologically illiterate people to their cuts, often with political motives.

(Paperback) ISBN-13: 978-1721756896, ISBN-10: 1721756892   £5.88 (US $7.75)
(Kindle) ASIN: **B07DZF75KD** $3.75 (US $5.00)

The Unintelligent Design Series

The Unintelligent Designer: Refuting the Intelligent Design Hoax

Showing why the superficial appearance of design in living things cannot be attributed to anything like an intelligent designer, as a counter to the politically motivated Intelligent Design movement.

(Hardcover) ISBN-13: 979-8513528463   £15.80 (US $20.00)
(Paperback) ISBN-10: 1723144215, ISBN-13: 978-1723144219   £11.20 (US $14.20)
(Kindle) ASIN B07G121BMK   £6.50 (US $7.50

The Malevolent Designer: Why Nature's God is not Good

Showing why, so much of nature could not have been designed by an intelligent all-loving creator.
Illustrated by Catherine Hounslow-Webber

(Hardcover) ISBN-13: 979-8511295442   £16.70 (US $21.00)
(Paperback) SBN-13; 979-8670361729   £11.20 (US $14.15)
(Kindle) ASIN: B08L9S8F5F   £6.75 (US $8.00)

Other Books By Rosa Rubicondior

Publish under the name Bill Hounslow – Oxfordshire Childhood series.

In The Blink of an Eye: Growing Up in Rural Oxfordshire

A frank recollections of life as feral children in the small North Oxfordshire hamlet of Fawler during the 1950s and 60s, on the brink of major change as we approached the television age and the final stages in the domestication of children was about to begin.
Additional material by Patricia Broome

(Hardcover) ISBN-13: 979-8511967400	£14.90 (US $19.00)
(Paperback) ISBN-10: 1545350787, ISBN-13: 978-1545350782	£8.95 (US $11.40)
(Kindle) ASIN: B06ZY8JZ92	£6.95 (US $8.95)

In The Blink of an Eye: Growing Up in Rural Oxfordshire Illustrated Edition.

Illustrated by Catherine Webber-Hounslow

(Hardcover) ISBN-13; 979-8364521361	£16.90 (US $21.60)
(Paperback): ISBN-13:979-8364503862	£10.55 (US $13.50)
(Kindle) ASIN: B0BNCQG8CC	£7.50 (US $9.10)

A Goose for Christmas: Stories from an Oxfordshire Childhood

Slightly imaginative stories, based on real events and people, of childhood adventures in the North Oxfordshire hamlet of Fawler in the 1950s during the post-war austerity, before television, when the children had only what they could get from the woods and fields around them.
Illustrated by Catherine Webber-Hounslow

(Hardcover) ISBN-13: 979-8511907482	£14.90 (US $19.00)
(Paperback) ISBN-13: 978-1981708925, ISBN-10: 1981708928	£9.35 (US $11.90)
(Kindle) ASIN: B07GFJ85P8	£6.50 (US $8.25)

www.ingramcontent.com/pod-product-compliance
Lightning Source LLC
Chambersburg PA
CBHW071042240526
45471CB00014B/271